ASCE STANDARD

ANSI/ASCE/EWRI 12-13
ANSI/ASCE/EWRI 13-13
ANSI/ASCE/EWRI 14-13

American Society of Civil Engineers

Standard Guidelines for the Design, Installation, and Operation and Maintenance of Urban Subsurface Drainage

Three Complete Standards

Standard Guidelines for the Design of Urban Subsurface Drainage
ANSI/ASCE/EWRI 12-13

Standard Guidelines for the Installation of Urban Subsurface Drainage
ANSI/ASCE/EWRI 13-13

Standard Guidelines for the Operation and Maintenance of Urban Subsurface Drainage
ANSI/ASCE/EWRI 14-13

This document uses both the International System of Units (SI) and customary units.

ASCE
AMERICAN SOCIETY OF CIVIL ENGINEERS

Approved American National Standard
ANSI

EWRI
ENVIRONMENTAL & WATER RESOURCES INSTITUTE

Library of Congress Cataloging-in-Publication Data

Standard guidelines for the design, installation, and operation and maintenance of urban subsurface drainage : three complete standards.
pages cm
At head of title: American Society of Civil Engineers.
Includes bibliographical references and index.
ISBN 978-0-7844-1305-0 (print) – ISBN 978-0-7844-7793-9 (pdf)
1. Subsurface drainage–Standards–United States. 2. Sewerage–Standards–United States. 3. Storm sewers–Standards–United States. I. American Society of Civil Engineers. II. Title: American Society of Civil Engineers standard guidelines for the design, installation, operation, and maintenance of urban subsurface drainage.
TC970.S6828 2013
628'.2021873–dc23

2013026708

Published by American Society of Civil Engineers
1801 Alexander Bell Drive
Reston, Virginia 20191
www.asce.org/pubs

This standard was developed by a consensus standards development process which has been accredited by the American National Standards Institute (ANSI). Accreditation by ANSI, a voluntary accreditation body representing public and private sector standards development organizations in the United States and abroad, signifies that the standards development process used by ASCE has met the ANSI requirements for openness, balance, consensus, and due process.

ISBN 978-0-7844-1305-0 (print)
ISBN 978-0-7844-7793-9 (PDF)
Manufactured in the United States of America.

STANDARDS

In 2006, the Board of Direction approved the revision to the ASCE Rules for Standards Committees to govern the writing and maintenance of standards developed by the Society. All such standards are developed by a consensus standards process managed by the Society's Codes and Standards Committee (CSC). The consensus process includes balloting by a balanced standards committee made up of Society members and nonmembers, balloting by the membership of the Society as a whole, and balloting by the public. All standards are updated or reaffirmed by the same process at intervals not exceeding five years.

The following standards have been issued:

ANSI/ASCE 1-82 N-725 Guideline for Design and Analysis of Nuclear Safety Related Earth Structures
ASCE/EWRI 2-06 Measurement of Oxygen Transfer in Clean Water
ANSI/ASCE 3-91 Standard for the Structural Design of Composite Slabs and ANSI/ASCE 9-91 Standard Practice for the Construction and Inspection of Composite Slabs
ASCE 4-98 Seismic Analysis of Safety-Related Nuclear Structures
Building Code Requirements for Masonry Structures (ACI 530-02/ASCE 5-02/TMS 402-02) and Specifications for Masonry Structures (ACI 530.1-02/ASCE 6-02/TMS 602-02)
ASCE/SEI 7-10 Minimum Design Loads for Buildings and Other Structures
SEI/ASCE 8-02 Standard Specification for the Design of Cold-Formed Stainless Steel Structural Members
ANSI/ASCE 9-91 listed with ASCE 3-91
ASCE 10-97 Design of Latticed Steel Transmission Structures
SEI/ASCE 11-99 Guideline for Structural Condition Assessment of Existing Buildings
ANSI/ASCE/EWRI 12-13 Standard Guidelines for the Design of Urban Subsurface Drainage
ANSI/ASCE/EWRI 13-13 Standard Guidelines for the Installation of Urban Subsurface Drainage
ANSI/ASCE/EWRI 14-13 Standard Guidelines for the Operation and Maintenance of Urban Subsurface Drainage
ASCE 15-98 Standard Practice for Direct Design of Buried Precast Concrete Pipe Using Standard Installations (SIDD)
ASCE 16-95 Standard for Load Resistance Factor Design (LRFD) of Engineered Wood Construction
ASCE 17-96 Air-Supported Structures
ASCE 18-96 Standard Guidelines for In-Process Oxygen Transfer Testing
ASCE 19-10 Structural Applications of Steel Cables for Buildings
ASCE 20-96 Standard Guidelines for the Design and Installation of Pile Foundations
ANSI/ASCE/T&DI 21-13 Automated People Mover Standards
SEI/ASCE 23-97 Specification for Structural Steel Beams with Web Openings
ASCE/SEI 24-05 Flood Resistant Design and Construction
ASCE/SEI 25-06 Earthquake-Actuated Automatic Gas Shutoff Devices
ASCE 26-97 Standard Practice for Design of Buried Precast Concrete Box Sections
ASCE 27-00 Standard Practice for Direct Design of Precast Concrete Pipe for Jacking in Trenchless Construction
ASCE 28-00 Standard Practice for Direct Design of Precast Concrete Box Sections for Jacking in Trenchless Construction
ASCE/SEI/SFPE 29-05 Standard Calculation Methods for Structural Fire Protection
SEI/ASCE 30-00 Guideline for Condition Assessment of the Building Envelope
SEI/ASCE 31-03 Seismic Evaluation of Existing Buildings
SEI/ASCE 32-01 Design and Construction of Frost-Protected Shallow Foundations
EWRI/ASCE 33-09 Comprehensive Transboundary International Water Quality Management Agreement
EWRI/ASCE 34-01 Standard Guidelines for Artificial Recharge of Ground Water
EWRI/ASCE 35-01 Guidelines for Quality Assurance of Installed Fine-Pore Aeration Equipment
CI/ASCE 36-01 Standard Construction Guidelines for Microtunneling
SEI/ASCE 37-02 Design Loads on Structures during Construction
CI/ASCE 38-02 Standard Guideline for the Collection and Depiction of Existing Subsurface Utility Data
EWRI/ASCE 39-03 Standard Practice for the Design and Operation of Hail Suppression Projects
ASCE/EWRI 40-03 Regulated Riparian Model Water Code
ASCE/SEI 41-06 Seismic Rehabilitation of Existing Buildings
ASCE/EWRI 42-04 Standard Practice for the Design and Operation of Precipitation Enhancement Projects
ASCE/SEI 43-05 Seismic Design Criteria for Structures, Systems, and Components in Nuclear Facilities
ASCE/EWRI 44-05 Standard Practice for the Design and Operation of Supercooled Fog Dispersal Projects
ASCE/EWRI 45-05 Standard Guidelines for the Design of Urban Stormwater Systems
ASCE/EWRI 46-05 Standard Guidelines for the Installation of Urban Stormwater Systems
ASCE/EWRI 47-05 Standard Guidelines for the Operation and Maintenance of Urban Stormwater Systems
ASCE/SEI 48-11 Design of Steel Transmission Pole Structures
ASCE/SEI 49-07 Wind Tunnel Testing for Buildings and Other Structures
ASCE/EWRI 50-08 Standard Guideline for Fitting Saturated Hydraulic Conductivity Using Probability Density Functions
ASCE/EWRI 51-08 Standard Guideline for Calculating the Effective Saturated Hydraulic Conductivity
ASCE/SEI 52-10 Design of Fiberglass-Reinforced Plastic (FRP) Stacks
ASCE/G-I 53-10 Compaction Grouting Consensus Guide
ASCE/EWRI 54-10 Standard Guideline for Geostatistical Estimation and Block-Averaging of Homogeneous and Isotropic Saturated Hydraulic Conductivity
ASCE/SEI 55-10 Tensile Membrane Structures
ANSI/ASCE/EWRI 56-10 Guidelines for the Physical Security of Water Utilities
ANSI/ASCE/EWRI 57-10 Guidelines for the Physical Security of Wastewater/Stormwater Utilities
ASCE/T&DI/ICPI 58-10 Structural Design of Interlocking Concrete Pavement for Municipal Streets and Roadways
ASCE/SEI 59-11 Blast Protection of Buildings
ASCE/EWRI 60-12 Guidelines for Development of Effective Water Sharing Agreement

CONTENTS

American Society of Civil Engineers

Standard Guidelines for the Design of Urban Subsurface Drainage

CONTENTS

FOREWORD

The Board of Direction approved revisions to the ASCE Rules for Standards Committees to govern the writing and maintenance of standards developed by ASCE. All such standards are developed by a consensus standards process managed by the ASCE Codes and Standards Committee (CSC). The consensus process includes balloting by a balanced standards committee and reviewing during a public comment period. All standards are updated or reaffirmed by the same process at intervals between five and 10 years. Requests for formal interpretations shall be processed in accordance with Section 7 of *ASCE Rules for Standards Committees,* which are available at www.asce.org. Errata, addenda, supplements, and interpretations, if any, for these standard guidelines can also be found at www.asce.org.

The *Standard Guidelines for the Design of Urban Subsurface Drainage* is a companion to the *Standard Guidelines for the Installation of Urban Subsurface Drainage* and *Standard Guidelines for the Operation and Maintenance of Urban Subsurface Drainage.* These standard guidelines were developed by the Urban Drainage Standards Committee, which is responsible to the Environmental and Water Resources Institute of the American Society of Civil Engineers.

The provisions of this document are written in permissive language and, as such, offer the user a series of options or instructions but do not prescribe a specific course of action. Significant judgment is left to the user of this document.

These standard guidelines may involve hazardous materials, operations, and equipment. These standard guidelines do not purport to address the safety problems associated with its application. It is the responsibility of whoever uses these standard guidelines to establish appropriate safety and health practices and to determine the applicability of regulatory and nonregulatory limitations.

These standard guidelines have been prepared in accordance with recognized engineering principles and should not be used without the user's competent knowledge for a specific application. The publication of these standard guidelines by ASCE is not intended to warrant that the information contained therein is suitable for any general or specific use, and ASCE takes no position respecting the validity of patent rights. The user is advised that the determination of patent rights or risk of infringement is entirely his or her own responsibility.

ACKNOWLEDGMENTS

The American Society of Civil Engineers (ASCE) acknowledges the work of the Urban Drainage Standards Committee of the Environmental and Water Resources Institute of ASCE (EWRI of ASCE).

This group comprises individuals from many backgrounds, including consulting engineering, research, the construction industry, education, and government. The individuals who serve on the Urban Drainage Standards Committee are

William Curtis Archdeacon, P.E., R.L.S., Past Chair
Kathlie Jeng-Bulloch, Ph.D., P.E., D.WRE, M.ASCE, Chair
Christopher B. Burke, Ph.D., P.E., D.WRE, F.ASCE
Jeffrey S. Glenn, P.E., D.WRE, CFM, F.ASCE
Robert S. Giurato
S. David Graber, P.E., F.ASCE, Vice Chair and Corresponding Editor
Jay M. Herskowitz, P.E., M.ASCE
Conrad G. Keyes, Jr., Ph.D., P.E., L.S., D.WRE, Dist.M.ASCE
James H. Lenhart, Jr., P.E., D.WRE, M.ASCE
Lawrence M. Magura, P.E., M.ASCE, Secretary
Garvin J. Pederson, P.E., M.ASCE
Anthony N. Tafuri, P.E., D.WRE, M.ASCE
Kenneth E. Waite, P.E., M.ASCE
William J. Weaver, P.E., M.ASCE
Donald E. Woodward, P.E., M.ASCE
Bin Zhang, P.E.

The corresponding editor recognizes the following committee members who were particularly helpful in updating the standard guidelines: Jeffrey S. Glenn, Jay M. Herskowitz, Conrad G. Keyes, Jr., James H. Lenhart, and William J. Weaver.

CHAPTER 1
SCOPE

The intent of these standard guidelines is to present state-of-the-art design guidance for urban subsurface drainage in a logical order. It updates ASCE/EWRI 12-05 *Standard Guidelines for the Design of Urban Subsurface Drainage* with material developed within the past five years. The collection and conveyance of subsurface drainage waters are within the purview of these standard guidelines for applications such as airports; roads; other transportation systems; and industrial, commercial, residential, and recreational areas. Incidental surface water is considered.

These standard guidelines do not address agricultural drainage, landfills, recharge systems, detention ponds, conventional storm sewer design, or the use of injection systems.

Customary units and standard international (SI) units are used throughout this document.

1.1 APPLICABLE STANDARDS

The following standards are available from the offices of the cited organization: American Association of State Highway Officials (AASHTO) in Washington DC; American National Standards Institute/American Water Works Association documents from AWWA in Denver, Colorado; American Society of Civil Engineers (ASCE) in Reston, Virginia; and ASTM International (ASTM) in West Conshohocken, Pennsylvania. The standards are mentioned in these guidelines in the sections where they are applicable.

The ASTM standard and comparable AASHTO standard for a product are frequently identical; however, there may be some differences, especially when AASHTO standards lag behind ASTM standard revisions. If a separate metric edition of a standard exists, its designation includes the letter M (e.g., C444M).

AASHTO, *Standard Specifications for Highway Bridges*, HB-17, 17th Ed., 2002.

AASHTO, *Standard Specification for Fine Aggregate for Hydraulic Cement Concret*e, M6, 2008.

AASHTO, *Standard Specification for Classification of Soils and Soil-Aggregate Mixtures for Highway Construction Purposes*, M145-91, 2004.

AASHTO, *Standard Specification for Bituminous Coated Corrugated Metal Culvert Pipe and Pipe*, M190-04, 2004.

AASHTO, *Standard Practice for Corrugated Polyethylene Drainage Pipe*, M252-09, 2009.

AASHTO, *Standard Specification for Precast Reinforced Concrete Box Sections for Culverts, Storm Drains, and Sewers*, M259/M259M-00, 2004.

AASHTO, *Standard Specification for Precast Reinforced Concrete Box Sections for Culverts, Storm Drains, and Sewers with Less Than 2 Feet (0.6 m) of Cover Subjected to Highway Loading*, M273/M273M-00, 2004.

AASHTO, *Standard Specification for Steel Sheet, Aluminum Coated (Type 2), for Corrugated Steel Pipe*, M274-87, 2004.

AASHTO, *Standard Practice for Corrugated Polyethylene Pipe 300- to 1500-mm (12- to 60-in.)*, M294-10, 2010.

AASHTO, *Standard Specification for Steel-Reinforced Polyethylene (PE) Ribbed Pipe, 300- to 900-mm (12- to 36-in.) Diameter*, MP 20-10, 2010.

American National Standards Institute/American Water Works Association (ANSI/AWWA) C300-04: *Reinforced Concrete Pressure Pipe, Steel Cylinder Type*, 2004.

ANSI/AWWA C301-07: *Prestressed Concrete Pressure Pipe, Steel Cylinder Type for Water and Other Liquids*, 2007.

ANSI/AWWA C302-04: *Reinforced Concrete Pressure Pipe, Noncylinder Type*, 2004.

ANSI/AWWA C303-08: *Concrete Pressure Pipe, Bar-Wrapped, Steel-Cylinder Type*, 2009.

ANSI/AWWA C900-07: *Polyvinyl Chloride (PVC) Pressure Pipe and Fabricated Fittings, 4 in. Through 12 in. (100-mm Through 300-mm), for Water Transmission and Distribution*, 2007.

ANSI/AWWA C905-10: *Polyvinyl Chloride (PVC) Pressure Pipe and Fabricated Fittings, 14 in. Through 48 in. (350-mm Through 1200-mm), for Water Transmission and Distribution*, 2010.

ASCE, *Standard Practice for Direct Design of Buried Precast Concrete Pipe Using Standard Installations*, Standard 15-98, 1998.

ASCE, *Standard Practice for Direct Design of Buried Precast Concrete Box Sections*, Standard 26-97, 1997.

ASCE, *Standard Practice for Direct Design of Precast Concrete Pipe for Jacking in Trenchless Construction*, Standard 27-00, 2000.

ASCE, *Standard Practice for Direct Design of Precast Concrete Box Sections for Jacking in Trenchless Construction*, Standard 28-00, 2000.

ASCE, *Comprehensive Transboundary Water Quality Management Agreement*, Standard 33-09, 2009.

ASTM, *Standard Practice for Structural Design of Corrugated Steel Pipe, Pipe-Arches, and Arches for Storm and Sanitary Sewers and Other Buried Appurtenances*, A796/A796M-10, 2010.

ASTM, *Standard Specification for Post-Applied Coatings, Pavings, and Linings for Corrugated Steel Sewer and Drainage Pipe*, A849-10, 2010.

ASTM, *Standard Specification for Clay Drain Tile and Perforated Clay Drain Tile*, C4-04, 2009.

ASTM, *Standard Specification for Installing Vitrified Clay Pipe Lines*, C12-09, 2009.

ASTM, *Standard Specification for Nonreinforced Concrete Sewer, Storm Drain, and Culvert Pipe*, C14-07/C14M-07, 2007.

ASTM, *Standard Specification for Concrete Aggregates*, C33/C33M-13, 2013.
ASTM, *Standard Specification for Concrete Pipe for Irrigation or Drainage*, C118-05a/C118M-05a, 2005.
ASTM, *Standard Test Method for Sieve Analysis of Fine and Coarse Aggregates*, C136-06, 2006.
ASTM, *Standard Specification for Reinforced Concrete Low-Head Pressure Pipe*, C361-08/C361M-08, 2008.
ASTM, *Standard Specification for Precast Reinforced Concrete Manhole Sections*, C478/C478M-09, 2009.
ASTM, *Standard Specification for Nonreinforced Concrete Irrigation Pipe with Rubber Gasket Joints*, C505-05a/C505M-05a, 2005.
ASTM, *Standard Specification for Vitrified Clay Pipe, Extra Strength, Standard Strength, and Perforated*, C700-09, 2009.
ASTM, *Standard Specification for Nonreinforced Concrete Specified Strength Culvert, Storm Drain, and Sewer Pipe*, C985-04/C985M-04, 2010.
ASTM, *Standard Specification for Precast Reinforced Concrete Monolithic Box Sections for Culverts, Storm Drains, and Sewers*, C1433/C1433M-10, 2010.
ASTM, *Standard Test Method for Compressive Properties of Rigid Cellular Plastics*, D1621-10, 2010.
ASTM, *Standard Test Method for Poly (Vinyl Chloride) (PVC) Plastic Pipe*, Schedules 40, 80, 120, D1785-06, 2006.
ASTM, *Standard Test Method for Polyethylene (PE) Plastic Pipe (SIDR-PR) Based on Controlled Inside Diameter*, D2239-03, 2003.
ASTM, *Standard Test Method for Poly (Vinyl Chloride) (PVC) Pressure Rated-Pipe (SDR Series)*, D2241-09, 2009.
ASTM, *Standard Practice for Underground Installation of Thermoplastic Pipe for Sewers and Other Gravity-Flow Applications*, D2321-09, 2009.
ASTM, *Standard Test Method for Permeability of Granular Soils (Constant Head)*, D2434, 2006.
ASTM, *Standard Practice for Classification of Soils for Engineering Purposes (Unified Soil Classification System)*, D2487, 2010.
ASTM, *Standard Specification for Acrylonitrile-Butadiene-Styrene (ABS) and Poly (Vinyl Chloride) (PVC) Composite Sewer Piping*, D2680-01, 2009.
ASTM, *Standard Specification for Poly (Vinyl Chloride) (PVC) Sewer Pipe and Fittings*, D2729-03, 2003.
ASTM, *Standard Specification for Acrylonitrile-Butadiene-Styrene (ABS) Sewer Pipe and Fittings*, D2751-05, 2005.
ASTM, *Standard Specification for Type PSM Poly (Vinyl Chloride) (PVC) Sewer Pipe and Fittings*, D3034-08, 2008.
ASTM, *Standard Test Method for Polyethylene (PE) Plastic Pipe (DP-PR) Based on Controlled Outside Diameter*, D3035-10, 2010.
ASTM, *Standard Test Method for Tensile Properties of Geotextiles by the Wide-Width Strip Method*, D4595-09, 2009.
ASTM, *Standard Test Method for Determining the (In-plane) Flow Rate per Unit Width and Hydraulic Transmissivity of a Geosynthetic Using a Constant Head*, D4716-08, 2013.
ASTM, *Standard Test Method for Determining Apparent Opening Size of a Geotextile*, D4751-04, 2004.
ASTM, *Standard Practice for Determining the Specification Conformance of Geosynthetics*, D4759-11, 2011.
ASTM, *Standard Test Method for Breaking Strength and Elongation of Textile Fabrics (Grab Test)*, D5034-09, 2009.
ASTM, *Standard Test Method for Measuring the Nominal Thickness of Geosynthetics*, D5199-11, 2011.
ASTM, *Standard Test Method for Evaluating the Unconfined Tension Creep and Creep Rupture Behavior of Geosynthetics*, D5262-07, 2007.
ASTM, *Standard Practice for Installation of Geocomposite Pavement Drains*, D6088-06, 2006.
ASTM, *Standard Test Method for Vertical Compression of Geocomposite Pavement Panel Drains*, D6244-06, 2006.
ASTM, *Standard Test Method for Determining Short-Term Compression Behavior of Geosynthetics*, D6364, 2006.
ASTM, *Standard Specification for Circular-Knit Geotextile for Use in Subsurface Drainage Applications*, D6707-06, 2006.
ASTM, *Standard Guide for Selection of Test Methods for Prefabricated Vertical Drains (PVD)*, D6917-03, 2007.
ASTM, *Standard Test Method for Testing Vertical Strip Drains in the Crimped Condition*, D6918-09, 2009.
ASTM, *Standard Specification for Geocomposites for Pavement Edge Drains and Other High-Flow Applications*, D7001-06, 2006.
ASTM, *Standard Test Method for Determining the Bond Strength (Ply Adhesion) of Geocomposites*, D7005-03, 2008.
ASTM, *Standard Guide for Acceptance Testing Requirements for Geonets and Geonet Drainage Geocomposites*, D7273-08, 2008.
ASTM, *Standard Test Method for Vertical Strip Drains Using a Large Scale Consolidation Test*, D7498-09, 2009.
ASTM, *Standard Practice for Corrugated Polyethylene (PE) Pipe and Fittings*, F405-05, 2005.
ASTM, *Standard Practice for Large Diameter Corrugated Polyethylene Pipe and Fittings*, F667-06, 2006.
ASTM, *Standard Specification for Poly (Vinyl Chloride) (PVC) Large-Diameter Plastic Gravity Sewer Pipe and Fittings*, F679-08, 2008.
ASTM, *Standard Specification for Smooth-Wall Poly (Vinyl Chloride) (PVC) Plastic Underdrain Systems for Highway, Airport, and Similar Drainage*, F758-95, 2007.
ASTM, *Standard Specification for Poly (Vinyl Chloride) (PVC) Profile Gravity Sewer Pipe and Fittings Based on Controlled Inside Diameter*, F794-03, 2009.
ASTM, *Standard Practice for Smoothwall Polyethylene (PE) Pipe for Use in Drainage and Waste Disposal Absorption Fields*, F810-07, 2007.
ASTM, *Standard Practice for Polyethylene (PE) Large Diameter Profile Wall Sewer and Drain Pipe*, F894-07, 2007.
ASTM, *Standard Specification for Poly (Vinyl Chloride) (PVC) Corrugated Sewer Pipe with a Smooth Interior and Fittings*, F949-10, 2010.
ASTM, *Standard Specification for Polypropylene (PP) Corrugated Wall Stormwater Collection Chambers*, F2418-09a, 2009.
ASTM, *Specification for Steel Reinforced Thermoplastic Ribbed Pipe and Fittings for Non-Pressure Drainage and Sewerage*, F2562/F2562M-08, 2008.
ASTM/AASHTO, *Standard Specification for Corrugated Steel Pipe, Metallic-Coated for Sewers and Drains*, ASTM A760/A760M-10, ASTM, 2010/AASHTO M36-03, AASHTO, 2003.
ASTM/AASHTO, *Standard Specification for Corrugated Steel Structural Plate, Zinc-Coated, for Field-Bolted Pipe, Pipe Arches, and Arches*, ASTM A761/A761M-04, ASTM, 2009/AASHTO M167M/M167-09, AASHTO, 2009.
ASTM/AASHTO, *Standard Specification for Corrugated Steel Pipe, Polymer Precoated for Sewers and Drains*,

ASTM A762/A762M-08, ASTM, 2008/AASHTO M245-00, AASHTO, 2004.
ASTM/AASHTO, *Standard Specification for Corrugated Aluminum Pipe for Sewers and Drains*, ASTM B745/B745M-97, ASTM, 2005/AASHTO M196-92, AASHTO, 2004.
ASTM/AASHTO, *Standard Specification for Reinforced Concrete Culvert, Storm Drain, and Sewer Pipe*, ASTM C76/C76M-11, ASTM, 2011/AASHTO M170-10/M170M-09, AASHTO, 2010.
ASTM/AASHTO, *Standard Specification for Concrete Drain Tile*, ASTM C412/C412M-05a, ASTM, 2005/AASHTO M178/M178M—07, AASHTO, 2007.
ASTM/AASHTO, *Standard Specification for Perforated Concrete Pipe*, ASTM C444/C444M-03, AASHTO, 2009/AASHTO M175/M175M-05, AASHTO, 2005.
ASTM/AASHTO, *Standard Specification for Reinforced Concrete Arch Culvert, Storm Drain, and Sewer Pipe*, ASTM C506/C506M-11, ASTM, 2011/AASHTO M206/M206M-10, AASHTO, 2010.
ASTM/AASHTO, *Standard Specification for Reinforced Concrete Elliptical Culvert, Storm Drain, and Sewer Pipe*, C507/C507M-10b, ASTM, 2010/AASHTO M207/M207M-10, AASHTO, 2010.
ASTM/AASHTO, *Standard Specification for Porous Concrete Pipe*, ASTM C654/C654M-05a, ASTM, 2005/AASHTO M176/M176M—07, AASHTO, 2007.
ASTM/AASHTO, *Standard Specification for Reinforced Concrete D-Load Culvert, Storm Drain, and Sewer Pipe*, ASTM C655/C655M-09, ASTM, 2012B/AASHTO M242/M242M-12, AASHTO, 2012.
ASTM/AASHTO, *Standard Specification for Acrylonitrile-Butadiene-Styrene (ABS) and Poly (Vinyl Chloride) (PVC) Composite Sewer Piping*, ASTM D2680-01, 2009/AASHTO M264-03, AASHTO, 2003.

CHAPTER 2
DEFINITIONS

2.1 GENERAL

This chapter defines specific terms used in these standard guidelines. The references listed in Chapter 10 may be useful to augment understanding of the terms in these guidelines.

2.2 TERMS

AOS—Apparent opening size of geotextiles.

Aquifer—Water-bearing stratum of permeable rock, sand, or gravel.

Barrier layer—Stratum with a hydraulic conductivity less than 10% of the weighted permeability of all the strata above it. Also referred to as the relative barrier.

Base drainage system—Permeable drainage blanket under a roadway pavement system.

Bedding—Granular material placed around subsurface drains to provide a structural support for the drain.

BTEX—An initialism for benzene, toluene, ethylbenzene, and xylenes, volatile compounds found in petroleum derivatives such as gasoline. These compounds are serious contaminants of soil and groundwater in some locations. They can have harmful effects on the central nervous system and cause other health issues.

Chimney drain—Subsurface interceptor drain frequently used in dams, embankments, and similar construction to control seepage within the earthen structure. Chimney drains are constructed in near-vertical orientation and discharge to outlets at lower elevations.

Colloidal fines—Clay particles smaller than two microns.

Consolidation drains—Wick drains that allow hydrostatic pressure to be relieved, thus allowing material to consolidate.

Drain envelope—Generic name for materials placed on or around a drainage product, regardless of whether used for mechanical support or hydraulic purposes (hydraulic envelope) or to stabilize surrounding soil material (filter envelope). Natural granular materials can also be used to improve bedding and backfill conditions.

Drainable water—Water that readily drains from soil under the influence of gravity. Also known as drainable porosity. Also see 6.5.2.

Evapotranspiration—Combined process of soil moisture evaporation and transpiration from plants.

Filter envelope—Permeable material placed around a drainage product to stabilize the structure of the surrounding soil material. A filter envelope may initially allow some fines and colloidal material to pass through it and into the drain.

Frost action—Movement of soil caused by freezing and thawing of soil moisture.

Geocomposite—Geosynthetic materials comprised of different combinations of geotextiles, geomembranes, geonets, and other materials for collecting and transporting water while maintaining soil stability.

Geology—Natural subsurface soil and rock formations.

Geomembrane—Sheet material intended to form an impervious barrier.

Geonet—A geosynthetic material consisting of integrally connected parallel sets of ribs overlying similar sets at various angles for in-plane drainage.

Geosynthetic—Synthetic material or structure used as an integral part of a project, structure, or system. Within this category are subsurface drainage and water control materials such as geomembranes, geotextiles, geonets, and geocomposites.

Geotextile—Woven or nonwoven engineered fabric intended to allow the passage of water and limited soil particles.

High column edge drains—Composite drain systems in which the height of the drain system is greater than the width.

Hydraulic barrier layer—Soil stratum with permeability less than 10% of the soil weighted permeability of the strata above it.

Hydraulic conductivity—A measure of the rate at which water moves in a soil or aggregate. When Darcy's Law applies, it is equal to the face velocity (the rate of flow divided by the corresponding total cross-sectional area) divided by the hydraulic gradient.

Hydraulic envelope—Permeable material placed around a drainage product to improve flow conditions in the area immediately adjacent to the drain.

Hydraulic gradient— In porous media, the hydraulic gradient is the rate of change in the sum of pressure head and elevation head (that sum being referred to as piezometric head) per unit distance. In an open channel, the hydraulic gradient equals the slope of the water surface.

Hydrology—Study of the movement of water in nature.

Infiltration—Passage of water into the soil.

Longitudinal drainage system—Drainage system parallel to a pavement system.

Perched water table—Localized condition of free water held in a pervious stratum because of an underlying impervious stratum.

MTBE—An initialism for methyl tert-butyl ether. This compound has been used, although decreasingly in the United States, as a gasoline additive to raise the octane number. It has been found to easily pollute large quantities of groundwater when gasoline with MTBE is spilled or leaked, raising health concerns.

Percolation—Downward movement of water through soil due to gravity.

Permeability (coefficient of permeability)—Rate at which a fluid moves through porous media under unit hydraulic gradient. In this publication, the fluid is always water and the media are soil, rock, aggregate drain materials, and consolidation drain materials.

Permittivity—Measure of the ability of a geotextile to permit water flow perpendicular to its plane.

Phreatic surface—Upper surface of an unconfined body of groundwater.

Relief drain—Any product or construction that accelerates the removal of drainable subsurface water.

Seepage—Movement of drainable water through soil and rock.

Sink—Surface depression that allows surface drainage to enter the subsurface water system.

Soil texture—Relative proportions of sand, silt, and clay particles in a soil mass.

Subsurface water—All water beneath the ground or pavement surface. Also see 6.2.1 and 6.4.2.

TCE—An initialism for trichloroethylene, a chlorinated hydrocarbon used as an industrial solvent. TCE has seeped into groundwater and raised health concerns in many locations.

Transmissivity—Hydraulic conductivity divided by media thickness.

Transverse drainage system—Drainage system usually at some angle to a roadway.

Water table—Upper limit of free water in a saturated soil or underlying material.

CHAPTER 3
SITE ANALYSIS

3.1 GENERAL

Site analysis involves a thorough review of existing information on the site and its surrounding area. Additional studies normally are needed to fill voids in data required to prepare proper design and contract documents.

3.2 BASIC REQUIREMENTS

3.2.1 General. The basic information necessary for the design of all subsurface drainage facilities requires investigations of the following features: topography, geography, water table, geology, water source, soil characteristics, environmental factors, and physical constraints.

3.2.2 Topography. All features that could influence subsurface drain location, installation, or operation must be considered in the design phase. Topographic studies of the drainage area ascertain the runoff direction following rainfalls, with depressed or poorly drained areas having greater drainage needs. A general topographic map of the project site and its surrounding area is required at the preliminary design stage. Topographic features of the area need to be of sufficient resolution to determine water movement to and from the drainage area. Surface water overland and concentrated flows frequently result from catchments and watersheds contributing to the area of concern.

A detailed topographic survey that locates planimetric features such as trees, ponds, ditches, culverts and catch basins, buildings, roads, walks, overhead utilities, and surface components of underground utilities and establishes elevations of such is necessary to develop and complete the design. The topographic information is used to establish proposed grades for the drain lines and outlet location and to determine the necessity and locations of pump stations.

Additional research may be required because subsurface conflicts not apparent from topographic analysis or record information may be present. For example, the digging of test pits, installation of subsurface piezometers, or examination of basements and subsurface drainage appurtenances in urban and agricultural areas may be needed to ascertain conflicts or subsurface water movement.

3.2.3 Geography. Geographic considerations may influence subsurface drainage system design. Coastal areas, floodplains, uplands, glaciated areas, humid areas, arid areas, and many other geographically distinct areas have unique drainage characteristics that the designer must consider during the design process. The designer is expected to become familiar with these unique characteristics before proceeding with the design of subsurface drainage facilities. Frequently, vegetation can indicate drainage characteristics and groundwater conditions. Seasonal influences can also influence the analysis and design of drainage systems in many areas. Groundwater conditions can vary significantly from season to season and, in many cases, can vary within a given season.

3.2.4 Water Table. The site may require an evaluation of the underlying water table early in the design stage. The designer must know the type of water table (confined or unconfined), the depth of the water table below the surface, the direction and gradient of groundwater flow, and any seasonal fluctuations in elevation.

Information related to water table fluctuation throughout the year must be evaluated. An understanding of lag time for water table response after precipitation events and fluctuations related to well pumping in the vicinity of the proposed subsurface drainage site may be necessary. Water table characteristics are particularly essential when the surface of a water table is above or in proximity to the anticipated subsurface drain outlet. It is important that the receiving drain has adequate capacity to accept this subsurface discharge continuously, particularly through periods of wet weather. Hydrostatic heads from confined aquifers, perched water tables, heterogeneous soil, and disturbed soil are important and should be noted in the site analysis reports. Where flat gradients exist, the direction of flow can be influenced by the orientation of subsurface drains, but where steep gradients exist, the subsurface drains must be oriented so as to intercept the flow. Drains oriented perpendicular to natural flow are the most efficient. Efficiency drops off rapidly as the orientation of the drains becomes parallel to the direction of flow. In some areas, geologic conditions can cause complex groundwater flow paths that can be difficult to intercept systematically with a standard subsurface drainage system.

With this information, the designer can evaluate the impact the subsurface drain will have on the water table. The design of the drain must be adequate to maintain the water table below desired limits during the design event. The designer may perform a risk assessment that predicts the frequency with which the water table can be expected to rise above the desired limits.

3.2.5 Geology. Geology plays an important role in slope stability analysis in cases where subsurface drains are needed to stabilize slopes. Relative permeability and direction of flow in earth materials that may affect the behavior of subsurface drainage are critically important in subsurface drain design. Geologic structure and history of deposition are also significant factors that determine the type of subsurface drainage system that is appropriate to meet project goals.

Geologic features such as rock layers, tight soil barriers, seepage paths, trench wall stability, and potentials for subsidence and collapse must be investigated.

3.2.6 Water Source. Present and projected water sources into the subsurface drain include precipitation; irrigation water and/or landscape water application; possible floodwaters; canal, reservoir, or pond seepage; surface stormwater; building roof runoff; subgrade drains; leaking septic appurtenances; or mounding from treated wastewater. Seasonally high water tables can affect the design, efficiency, and operation of the subsurface drain and must be considered.

Surface water should not be introduced into a subsurface drainage system, if at all possible, to prevent the accumulation of debris or other deleterious matter that may cause plugging of the drains and increase maintenance costs. The most cost-efficient system in terms of life-cycle costs may include separate systems to collect and drain surface water and subsurface water. However, in many areas surface water infiltration is increasingly used to manage stormwater runoff. These stormwater management best management practices (BMPs) capture surface runoff from impervious surfaces to facilitate longer-term infiltration into beds of engineered soils and thus allow for deep infiltration into underlying soils or evapotranspiration between runoff events. Infiltration systems have also been used to direct runoff that is relatively low in suspended solids, such as from roof tops, into the ground. However, the potential pollution from roof tops should be considered (see Clark et al. 2008). These systems should include features that provide for periodic maintenance access and trap constituents that can clog the system.

3.2.7 Soil Characteristics

3.2.7.1 General. Many soil profiles are nonuniform and contain layers of various soil types or rock with varying properties. Within these strata, the movement of groundwater may vary greatly. Relatively pervious soil and rock formations (aquifers) may exist to influence water movement. An aquiclude (relatively impervious stratum) also affects movement of groundwater.

Permeable soil profiles may not drain freely because of impervious soils or structures near the outer regions; thus, the soil profile at the boundary of areas to be drained must be reviewed. In the soil profile study, the location of an impermeable barrier layer must be established. Subsurface drains must be placed above this layer. The efficiency of the drainage system relates to the distance between the drain invert and the impermeable layer. The barrier layer may be nothing more than a change in soil texture within the profile. In practice, the barrier layer is a stratum with permeability less than 10 to 20% of the weighted permeability of all layers above it.

Specific details on soil classification, strata and layers, permeability or hydraulic conductivity, soil-water chemistry, temperature, and area vegetation are important design considerations.

3.2.7.2 Soil Classification. Information on soil classification can be presented in accordance with the Unified Soil Classification System (Terzaghi, Peck, and Mesri 1996; ASTM D2487), USDA Textural Classification System (USBR 1993), or AASHTO Classification System (AASHTO M145). The primary concerns are the texture, structure, and depth of soil layers; barriers to water movement; apparent permeability (referring to the permeability in the most significant direction in the case of an anisotropic medium) and surface infiltration potential; and surface or shallow bedrock. Soil structural characteristics such as shrink-swell, cracking, subsidence, collapse, consolidation, surface sealing, and compaction potentials are also important. The USDA Textural Classification System (USBR 1993) has some preferred details for use in Chapter 5, but all systems named previously will help in determining classifications.

Infiltration rate can be properly determined as described in the U.S. Bureau of Reclamation's *Drainage Manual* (USBR 1993), Section 3-10.

3.2.7.3 Strata and Layers. The soil strata and layering require identification because they affect the water movement to and the installation of the subsurface drain lines. Depth, thickness, permeability, slope, and extent of the layers are required data. In historically agricultural and urban soils significant variances can exist in soil characteristics relating to mixing, imported soils, compaction, and abandoned trenches or subsurface drainage systems such as agricultural drainage tiles.

3.2.7.4 Hydraulic Conductivity and Permeability. These factors should be determined through appropriate tests. In situ saturated hydraulic conductivity tests are recommended. Several procedures are presented in USBR 1993. Laboratory permeability tests are useful as a screening tool but are not recommended for design of subsurface drains.

3.2.7.5 Soil-Water Chemistry. Salinity, corrosivity, and pH must be determined before selecting pipe and pump materials. Any substances present or projected at the site that could affect either material selection or disposal of the drain effluent must be identified. Substances such as iron pyrites, sodium, calcium, selenium, boron, arsenic, or iron may create disposal and/or material selection problems. Soils with unusually high sodium content are difficult or impossible to drain. Screening tests, including laboratory permeability, pH, settling volume, and electrical conductivity, are useful in detecting the probability of excessive sodium in soils.

3.2.7.6 Temperature. An understanding of the temperature range and fluctuations is important for evaluating potential frost action or freeze-thaw. Frost action occurs when moisture is in the freezing zone of the soil and is caused by the freezing and thawing of such moisture. Frost action may influence material selection and installation depth.

3.2.7.7 Vegetation. Trees and shrubs are particularly important design factors. Water-seeking vegetation, such as the willow tree, is critical in subsurface drain location because the roots tend to plug drains. Other vegetation and landscape details must be noted to provide information on rooting depths, which is important to drain depth and spacing.

3.2.8 Environmental Factors

3.2.8.1 General. Major environmental factors must be considered in the design of subsurface drainage systems to prevent adverse environmental effects to adjacent parcels of land, residents, and environmentally sensitive ecosystems. These factors include the following:

- Water quality
- Flooding
- Wetlands
- Principal or primary aquifers
- Hydrology

3.2.8.2 Water Quality. Design, construction, and operation of subsurface drainage systems and components must be performed such that nearby channels are not contaminated. Contamination occurs when subsurface water accumulates excessive minerals, pesticides, herbicides, nutrients, or other soluble organics, such as BTEX, MTBE, or TCE, as the water passes through the soil. Water tables crucial to agricultural farm applications and/or depended on as sources of potable water must be of high quality. Certain states of the United States regulate the possibilities of contamination through required permits for discharge. Such permit requirements must be adhered to in design, construction, and operation. Employing oil-water separation facilities in parking areas and roadways and the selective use of discharge points are some of the options a designer can use to mitigate possible contamination of water.

3.2.8.3 Flooding. In areas where projects divert a high-flow volume of water offsite, the designer should consider the possibility of flooding. Flooding is caused by overflowing streams and runoff from adjacent properties. Potential flooding must also be examined in situations where subsurface drainage is diverted into certain channels such as ditches, enclosed drainage networks, and/or streams. Sufficient capacity for the subsurface flow system should be provided in the channel during storms. An analysis of downstream channel conditions should be performed to evaluate flood hazards and assess the impact of flooding backflow into the subsurface drainage system. Potential flooding or drainage system backflow effects should be adequately considered.

3.2.8.4 Wetlands. Additional consideration is required for the ecological-environmental aspects of the site when applying artificial drainage to a wetland area.

Wetlands are classified by local, state, and federal governments. Most wetlands are considered to be environmentally sensitive ecosystems. The maintenance of these ecosystems is important. Many wetlands filter natural and artificial pollutants. Changes in the quality and/or quantity of subsurface drainage waters entering a wetland can adversely affect this sensitive filtering process and may cause detrimental effects to flora and fauna associated with the wetland. Influences of subsurface drainage on wetland hydrology and groundwater quality can also be important design factors. Fen and other unique and rare wetlands depend on chemical characteristics associated with groundwater seep hydrology sources. In these situations, the effect of changes to groundwater recharge areas with respect to water quantity and quality should be adequately considered.

Any project associated with a wetland will most likely require permits from local, state, and/or the federal government.

3.2.8.5 Principal or Primary Aquifers. These aquifers are often tapped as a main water source. The intended use of this water determines the necessity and amount of protection required. If a possibility exists for contamination of the aquifer, mitigation measures must be taken to prevent such contamination. Other design alternatives, including relocation of the system or a treatment and monitoring program of the subsurface discharge, may be necessary to remove the contamination potential.

3.2.8.6 Hydrology. Hydrology describes the movement of water. Development modifies the natural hydrologic cycle and generates an artificial water cycle. Although this has typically been an insignificant design factor, it has become a valid concern in recent years. The development of subsurface drainage systems should follow the natural hydrologic cycle as closely as possible. For example, if the natural cycle exists as rainfall that percolates into groundwater and then joins surface watercourses, the artificial cycle should mimic this movement. Not all hydrologic cycles are this simple and thus easily mimicked, but considering the natural or existing hydrologic cycle of the site in the design is important.

3.2.9 Physical Constraints. Most urban settings have constraints related to existing or planned utilities that must be considered in the design of subsurface drainage systems. Compatibility of proposed systems with existing drain systems is critical to any layout. The location of utilities may require special consideration in the design stage to accommodate pump stations, future development, and master planning. Additional physical constraints can be identified through a topographic survey, as discussed in 3.2.2.

CHAPTER 4
SYSTEM CONFIGURATION

4.1 GENERAL

An urban subsurface drainage system will include any or all of the following components: collection and conveyance conduits; special conduit bedding, filter, and backfill materials; outlet structures; and appurtenances.

4.2 COLLECTION SYSTEM TYPES

4.2.1 Pipes. Pipes in these guidelines are intended only for the collection and conveyance of water. Modern pipe materials include concrete, ductile iron, steel, plastic, and clay, and the pipe may have a solid, slotted, or perforated wall. New composite subsurface drainage systems and conduits are also available.

Subsurface drains are generally installed with the perforations down (ASTM C12, NCPI 2006, NCSPA 2008, ODOT 2010), below the top half and above the bottom sector or, for some pipe types and sizes, below the top half and above the full bottom quadrant of the pipe. Such placement of the perforations may reduce the entrance of fine soil solids, reduce clogging of perforations, improve hydraulic characteristics, keep low seepage flows within the pipe (and thus improve low-flow velocities), and provide greater pipe strength (Graber [2004] and references cited therein). Pipes are joined in various ways to provide soil-tight or watertight joints. Connector devices include tees, wyes, elbows, and adapters for different diameters and materials.

4.2.2 Geocomposites and Geonets. Geocomposites and geonets of similar or different materials may be used as collection devices to create an "in-plane" envelope for intercepting liquid flowing at right angles to the envelope. Geocomposites and geonets may also be used to convey liquid to an outlet. Geocomposites and geonets may be installed vertically (i.e., attached to the exterior of a structural foundation wall) or horizontally in a trench.

4.2.3 Geomembranes. Geomembranes may be used as a barrier to liquid flow to waterproof foundation walls, seal under pavements, cutoff fills, and line cuts (e.g., in earth dams, hazardous waste dumps, drainage ditches, and retaining walls). Geomembranes are manufactured in a variety of materials, including plastic, synthetic rubber, and asphaltic compounds.

4.2.4 Geotextiles. Geotextiles may be used as coverings and liners in several construction applications. In drainage applications, geotextiles serve as filters that pass water and colloidal fines while restricting soil migration.

Extra care should be taken in the design of systems utilizing geotextiles to prevent pore blockage by migrating fines. Fabric failures due to pore blockage are extremely difficult to remedy.

4.2.5 Aggregates. Aggregates may be used as filters or envelopes. Aggregates may be either sands and gravels or crushed stone. Certain artificial aggregates, such as crushed glass, are also used. Aggregates should be essentially free of sediment, foreign materials, and any contaminants of concern.

4.2.6 Wick Drains. Wick drains may be used as vertical drainage to abet the upward flow of water from underground sources and make it free flowing on the surface. Wick drains accelerate the consolidation of soft and compressible soils.

4.3 CONVEYANCE CONDUITS AND OUTLETS

Water collected in a drainage system is normally conveyed to a safe and adequate outlet, such as a natural outfall or storm drainage facility. Where gravity flow is not feasible, pumping is necessary. Most of the collection system types in 4.2 may also be used in conveyance systems, which have different design and installation requirements.

4.4 APPLICATIONS

This section includes, but is not limited to, the major applications of the collection system types described in 4.2. Combinations of these systems may be used in a drainage system. Geotextiles typically encapsulate aggregate to form a viable drainage system.

4.4.1 Foundation Drains. Foundation drains have application to buildings, bridges, dams, and retaining walls, where structural elements are involved and removal of water is needed. All collection system types have application in foundation drains. In many foundations, a subsurface drainage system using pipes, geonets, geocomposites, geomembranes, geotextiles, and/or aggregates is used. Geomembranes by themselves will keep a structure dry but do not remove the subsurface water. Some composite materials provide an impervious membrane with dimples that have attached geotextiles. This provides an open channel between the two for water to freely drain while preventing the seepage of water into dry foundations. The use of wick drains is limited to areas where the hydraulic head (soil pore pressure) is great enough to force subsurface water to the surface. A wick drain system can experience maintenance problems where surface grading is slight and freeze-thaw cycles are a reality. Wick drains have also been used to drain groundwater from steep bluff settings; however, care must be taken in areas where water-bearing soil layers are not uniform. In those areas, installing wick drains in a way that will effectively and predictably capture groundwater can be difficult. See Fig. 4-1 for typical applications of foundation drains.

4.4.2 Roads, Railroads, and Airports. Roads, railroads, and airports have pavement systems with significant length-to-width ratios. The types of drainage systems typically are longitudinal, transverse, and base.

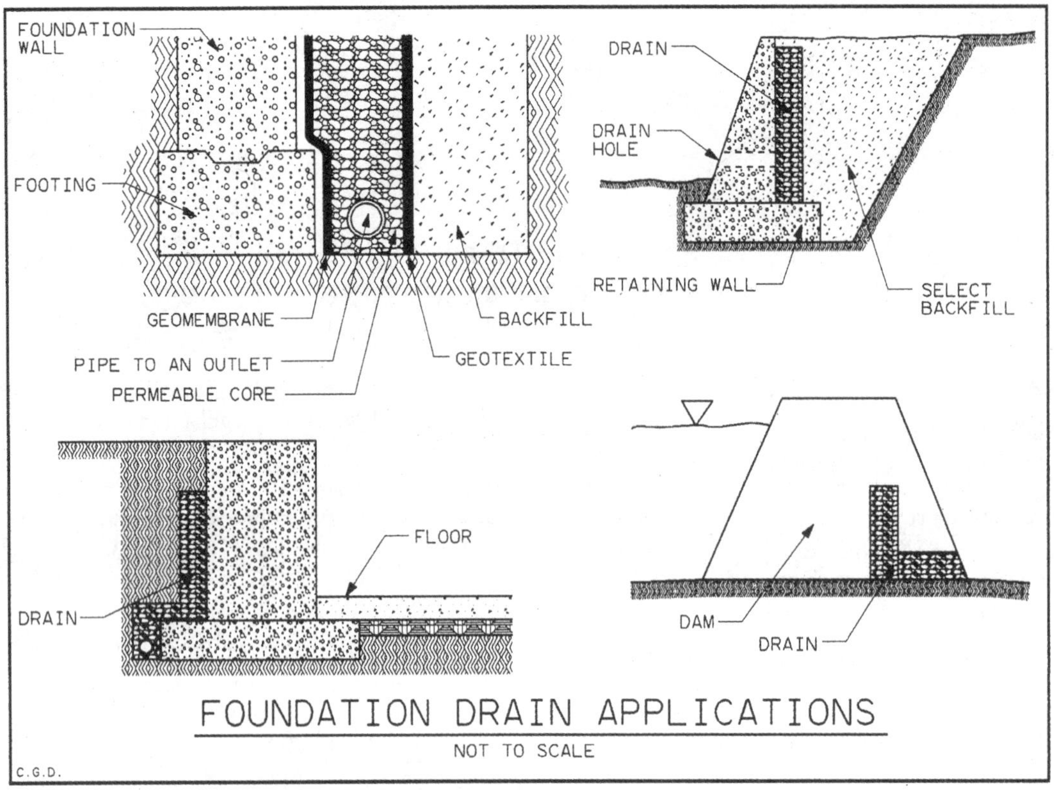

FIGURE 4-1. Foundation drain applications

4.4.2.1 Longitudinal Drainage System. A longitudinal system is essentially parallel to the pavement system. It is usually placed at a depth that allows the road section structural base to be gravity drained. Plastic perforated pipe systems, geotextiles, aggregates, structural geonets, and geocomposites are commonly used for longitudinal systems. When curb and gutter are present in a roadway system, discharge into a surface drainage system, frequently tied into the roadway catchbasins, is usually required. See Fig. 4-2 for a typical application of a longitudinal drainage system.

4.4.2.2 Transverse Drainage System. A transverse system drains across the pavement system. It is usually at right angles to the pavement but can be placed at any angle. A transverse drainage system consists of pipe, geotextile, aggregate, geonet, and/or geocomposite. The main need for this system is in sections where problems with groundwater are anticipated. This system should connect to the longitudinal subsurface system or the surface drainage system. Figures 4-3 and 4-4 show typical transverse drainage systems.

4.4.2.3 Base Drainage System. A base drainage system usually consists of a permeable drainage blanket under the entire pavement system. This system can also add to the structural integrity of the pavement. A properly graded aggregate is often sufficient for a base drainage system. Geotextiles are often used with an aggregate to provide separation and filtration or strength. A properly crowned roadway, railroad, taxiway, and runway subbase will enhance the drainage system. Figure 4-5 shows a typical base drainage system.

4.4.2.4 Permeable Pavements. Quite a few permeable unit paving systems are used in road and street applications around the United States and around the world. The Chicago green alley program (CDOT 2007) is the largest use of permeable paving in the United States. Also see 4.4.3.1.

4.4.3 Parking and Other Paved Areas. Parking and other paved areas such as parking lots and play areas have pavement systems where the length-to-width ratio approaches 1. The type of drainage needed varies with the paved surface. Stone and artificial turf are considered pervious surfaces. Ordinary concrete is considered to be a relatively impervious surface, though porous concrete is used to manage stormwater runoff. Asphalt, depending on design, may be pervious or impervious. In addition, many parking areas utilize porous pavers, grass pavement, and other manufactured products to reduce runoff from paved surfaces.

4.4.3.1 Pervious Facilities. These facilities (e.g. pervious parking lots) can be large with numerous low points where surface and subsurface drainage collect. Accumulated drainage is conveyed through the pervious surface to the subsurface collector system of pipes, geonets, geocomposites, and/or aggregates.

The more pervious or larger the facility, the greater are the rates and amounts of discharge to the subsurface system. Often these systems are constructed in tandem with a surface drainage system.

Geotextiles and geomembranes can be used depending on effects of the water table. The most important design consideration is to properly design the subbase to convey subsurface drainage into the pipe system. The complex flow characteristics of pervious systems require thorough study. Figure 4-6 illustrates a typical pervious system.

4.4.3.2 Impervious Facilities. A large impervious facility typically has numerous low points to collect surface and subsurface

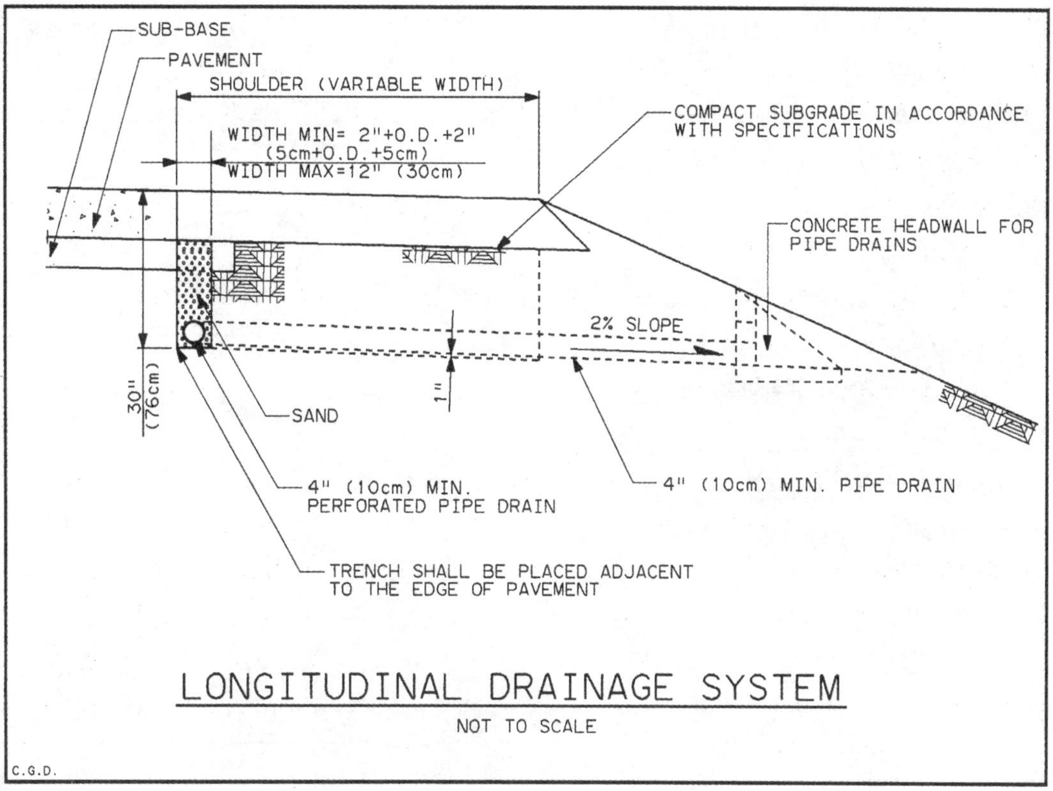

FIGURE 4-2. Longitudinal drainage system

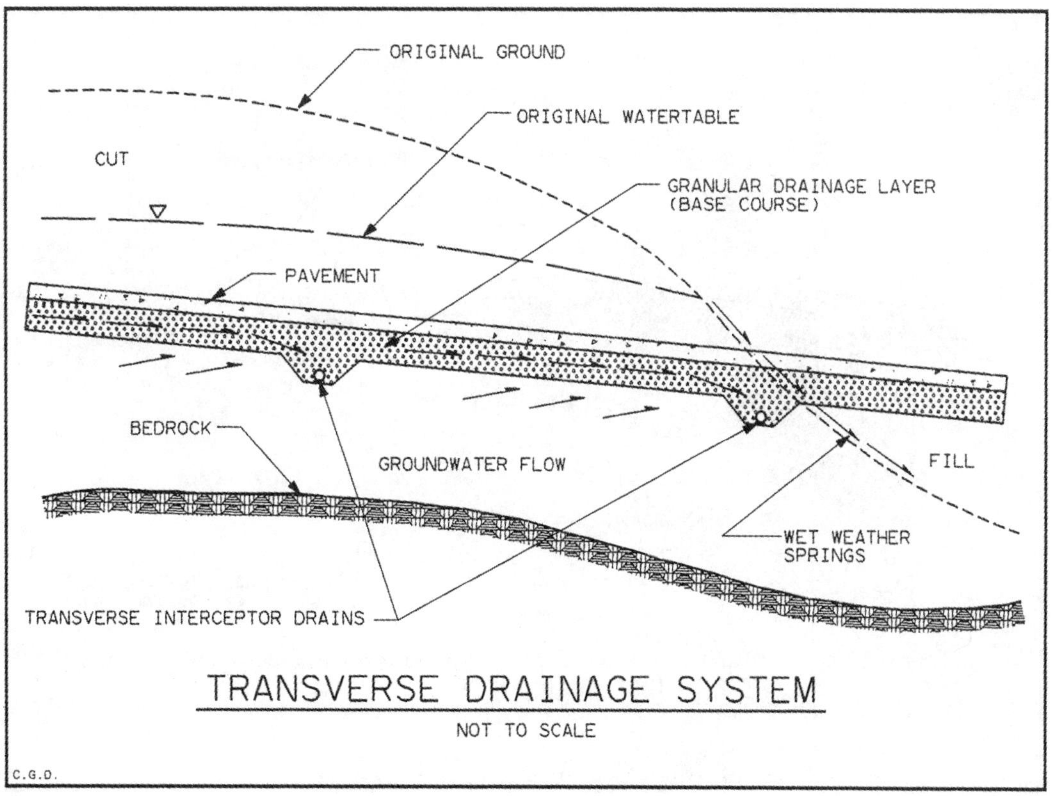

FIGURE 4-3. Transverse drainage system

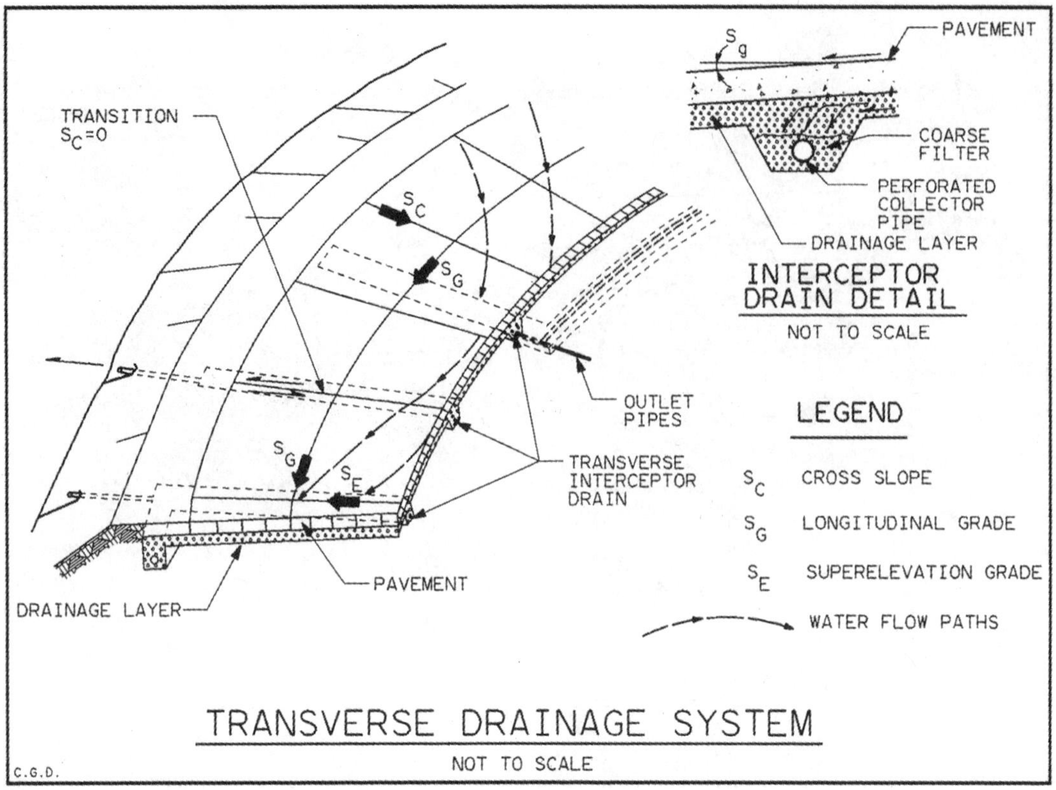

FIGURE 4-4. Transverse drainage system

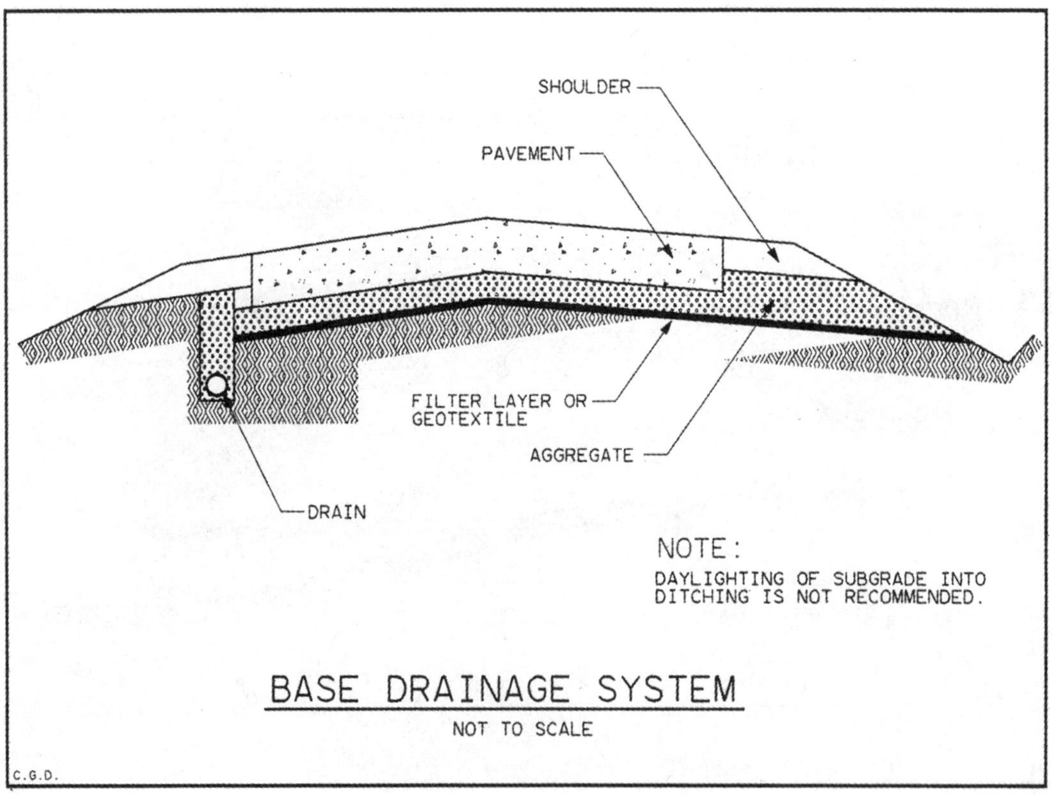

FIGURE 4-5. Base drainage system

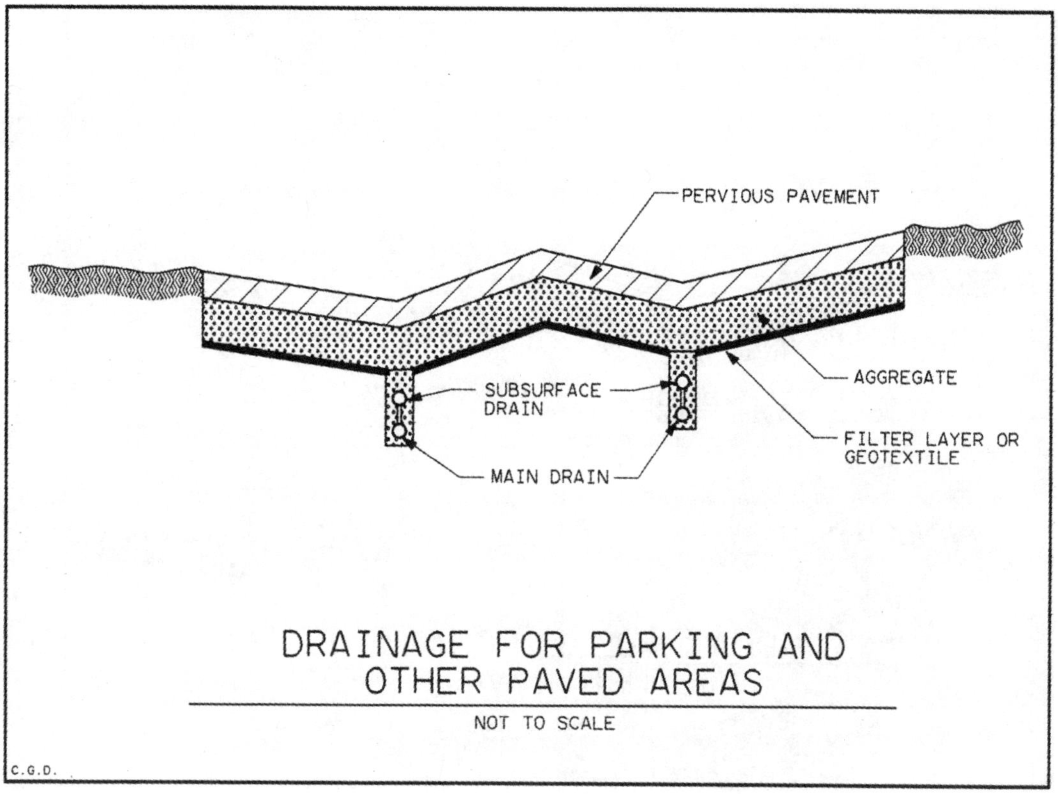

FIGURE 4-6. Drainage for parking and other paved areas (pervious)

drainage. However, the volumes of subsurface water are usually very small due to low rates of infiltration through the pavement. Perforated pipes, aggregates, geonets, geocomposites, and geotextiles are used in these types of facilities. Unless a rising water table is present, geomembranes are seldom used, because there is rarely a need to keep an area completely dry. Figure 4-7 illustrates a typical impervious system.

4.4.4 Recreational and Turf Areas. Recreational and turf areas, such as parks, golf courses, and athletic fields, are also used for detention or retention of surface water. Because of the relatively large size of these facilities, the cost of a subsurface drainage system can be significant. The system used should be designed for low maintenance and should take into account variations of the surface over time. All the collection systems are applicable in one form or another, although the importance of keeping these areas "dry" should be weighed before a system is designed. For example, a revenue-producing golf course must drain well to be back in service as soon as practicable after a storm ends. Keeping adjacent surface waterways free of debris can also be an important part of controlling groundwater levels. Figure 4-8 shows a typical application for a recreational and turf area.

4.4.5 Landscaped Areas. Landscaped areas are usually situated in urban areas and are typically small in relation to the overall development. Large landscape areas are usually part of an overall recreational area. The importance of keeping the areas "dry" must be weighed before a system is designed. For example, an area that has a large amount of foot traffic must be back in service soon after a storm ends. However, in many areas of the United States, landscaped areas are designed to pond and/or infiltrate stormwater runoff (Ferguson 1994). These areas are considered BMPs and can be stormwater planters, bioinfiltration facilities, or biofiltration facilities. Some facilities, such as swales and filter strips, serve as detention basins or convey water. The use of engineered soils is common to maintain infiltration rates on the order of 2 to 6 inches per hour or more.

All collection systems, except wick drains, are applicable. Wick drains are not used, because their resultant surface drainage runoff is unacceptable in an urban area. Geomembranes may be used in planter areas to keep foundations dry. Pipe systems may be used in tandem with city drainage systems. Geocomposites, geonets, geotextiles, and aggregates may be used to remove surface ponding. The amount of drainage provided should be related to the effect of ponding on the landscaped area.

The use of BMPs should be approached with regard to both their real-world efficacy and the need for long-term maintenance for continued effectiveness. Schneider and McCuen (2006, p. 278) note that "Although policies that require the use of BMPs are already in place, little research has addressed the effectiveness of BMPs. BMPs are being implemented even though their real-world effectiveness has not been fully assessed." Emerson and Traver (2008, p. 598) caution: "There have been previous surveys of infiltration BMPs that found high failure rates within a relatively short time frame....The conclusion of [one field inspection of such BMPs that were relatively new (≤2 years] was that 67% of the BMPs were not functioning as intended." Lee, Heaney, and Pack (2010, p. 237) state, "A number of storm-water controls that are commonly referred to as best management practices (BMPs) or sustainable urban drainage systems, have been introduced, but it is not simple to evaluate the benefits of storm-water management measures from those approaches." Lee, Heaney, and Pack (2010) provide related suggestions for improved evaluations.

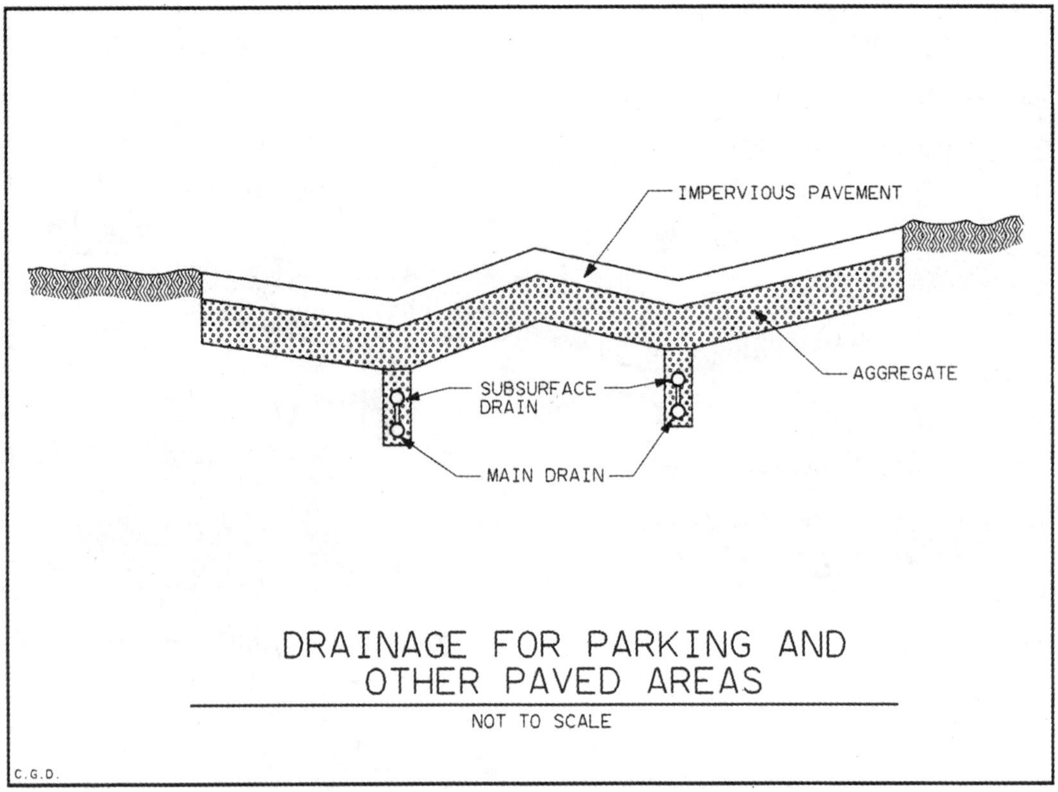

FIGURE 4-7. Drainage for parking and other paved areas (impervious)

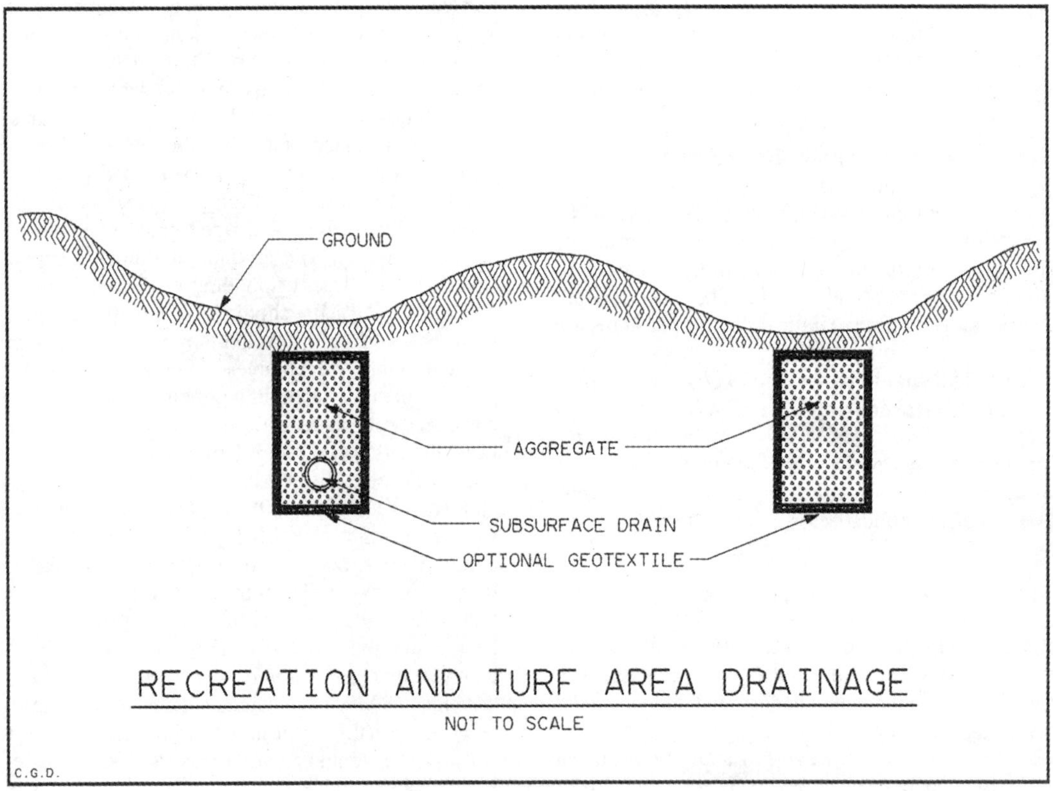

FIGURE 4-8. Recreation and turf area drainage

4.5 APPURTENANCES

An urban subsurface drainage system may include various appurtenances necessary for a complete and operational system, including such items as pumping and lift stations, vaults, manholes, and cleanouts.

4.5.1 Pumping and Lift Stations. Pumping stations and lift stations may be used in conveyance systems to transport water to a distant and higher discharge outlet. Pumping stations normally include pumps, piping, valves, ducts, vents, controls, electrical equipment, and accessories.

4.5.2 Vaults. Vaults may be used in any drainage system to house electrical or other equipment underground. Vaults normally include ducts, piping, valves, vents, and accessories in addition to the equipment being housed.

4.5.3 Manholes. Manholes may be used in conveyance systems to facilitate inspection and maintenance of the drainage pipe. In small-diameter pipe systems that cannot be entered by personnel, manholes are normally constructed at each change in grade, pipe size, or alignment, and at intervals for cleaning purposes. In pipe systems that can be entered by personnel, changes in alignment may be accomplished by curving the pipeline to eliminate the need for a manhole. Manholes may be constructed of concrete or other approved materials. Precast concrete or prefabricated manhole units are joined in various ways to provide soil-tight joints. Pipe-to-manhole connections should be soil-tight to prevent piping of backfill material into the manhole. The connection should also provide flexibility at the pipe-manhole interface. A watertight connection may be required in some installations. Manholes are normally capped with a metal casting with a removable lid or with a concrete slab that includes the metal casting and removable lid.

4.5.4 Cleanouts. Cleanouts may be used in conveyance systems to facilitate inspection and maintenance of drainage systems that cannot be entered by personnel and are not scheduled for frequent inspection and maintenance. Cleanouts are normally constructed at grade and alignment changes of approximately 45° or greater. Cleanouts are usually a wye section in the pipeline, with a removable stopper in the wye. Cleanouts in public rights of way are normally extended to the surface or to a point 6 to 12 inches (150 to 300 mm) below finished ground surface and plugged with a removable stopper. Cleanout wyes should be the same material as the main pipeline. Cleanout extensions may be of any approved pipe material.

CHAPTER 5
DRAIN ENVELOPES

5.1 GENERAL

Drain envelopes, as used here, is an inclusive term that encompasses any type of material placed on or around a drain for any reason. Refer to Chapter 2, Definitions, for more complete definitions of drain envelopes, hydraulic envelopes, and filter envelopes. The primary reasons for placing envelope materials around subsurface drains are

- To prevent excessive movement of soil particles into the drain,
- To provide material in the immediate vicinity of the drain that is more permeable than the surrounding soil,
- To provide bedding for the drain, and
- To stabilize the soil in which the drain is being placed.

5.2 DETERMINATION OF NEED

Envelope characteristics should be considered in most applications. This is true for all pipe types for structural purposes and to increase water flow rates and, possibly, water storage capacity. Refer to Table 5-1 for guidance.

Filters are required in soils with piping or migration potential. Generally, Table 5-1, from the Soil Conservation Service (now Natural Resources Conservation Service), provides good guidance based on soil classification. Other factors, such as perforation size in the pipe and velocity of flow, must also be considered. Filtration requirements take precedence over other considerations for envelope design needs.

5.3 DESIGN OF ENVELOPES

In selecting a proper filter medium, determining the gradation curve using wet-sieve analysis techniques of the in situ material to be drained is necessary. This should be done using actual site samples, because gradations for a given soil type can vary significantly. From these curves the following design calculations establish limits that are acceptable for filter material. The first criteria proposed by Terzaghi (U.S. Army Corps of Engineers 1941) for what he termed a filter are as follows:

- The particle diameter of the D_{15} size of the filter material should be at least four times as large as the diameter of the 15% size of the base material (D_{15}). (This makes the filter material roughly more than 10 times as pervious as the base material.)
- The 15% size of the filter material (D_{15}) should not be more than four times as large as the 85% size of the base material (D_{85}). (This prevents the fine particles of the base material from washing through the filter material.)

The D_{15} size is the particle diameter such that 15% of the material by weight is of a smaller diameter. Also, 85% of the base material, by weight, is smaller than the D_{85} size of the soil. In the case of drainage, the base material is the soil.

The Soil Conservation Service (USDA, SCS 1991) utilized the filter criteria by Terzaghi as outlined here, along with additional laboratory research (Sherard et al. 1984a, 1984b) and field experience in developing its sand gravel filter envelope criteria (USDA, SCS 1991). USDA, SCS (1991) recommends the following gradation limits:

- Upper limit of D_{100} is 38 mm (1.5 in.).
- Upper limit of D_{15} is the larger of 7 times D_{85} or 0.6 mm.
- Lower limit of D_{15} is the larger of 4 times D_{15} or 0.2 mm.
- Lower limit of D_5 is 0.075 mm (number 200 sieve).

Equation 5-1 and 5-2 provide an acceptable alternative that satisfies both filter and hydraulic envelope requirements when used in conjunction with the recommended Coefficient of Curvature C_c and the recommended Coefficient of Uniformity C_u (USBR 1993, ASCE 1998). The gradations shown are based on soil texture. The C_c must be greater than 1.0 and less than 3.0 for both gravels and sands, and the C_u must be greater than 4.0 for gravels and greater than 6.0 for sands, with:

$$C_u = \frac{D60}{D10} \quad (5\text{-}1)$$

and

$$C_c = \frac{(D30)^2}{(D10)(D60)} \quad (5\text{-}2)$$

where $D10$, $D30$, and $D60$ = diameter of particles passing the 10, 30, and 60% points on the envelope material gradation curve. Not more than 5% of the filter material should pass the #200 sieve or D_5 (filter) > 0.075 mm. This limit is to ensure the permeability of the filter.

The envelope material for the entire drain system should be designed to work with the most restrictive base soil that will be encountered.

5.3.1 Pipe Holes, Slots, and Joints. Envelope materials (or in situ soils if filters are omitted) must be coarse enough not to enter the pipe openings. The following equation should be used:

$$\frac{D_{85}\text{ (filter)}}{\text{maximum opening in pipe}} \geq 1 \quad (5\text{-}3)$$

5.3.2 Bedding Envelopes. Bedding envelopes may be used to provide proper bedding for sealed or nonperforated drainage pipe. The same design criteria are used for filter envelopes and hydraulic envelopes, but they may be more expensive than necessary if a well-graded coarse sand-gravel material will meet the need.

Table 5-1. Classification to Determine the Need for Drain Filters or Envelopes and Minimum Velocities in Drains

Unified Soil Classification	Soil Description	Filter Recommendation	Envelope Recommendation	Recommendations for Minimum Drain Velocity
SP (fine)	Poorly graded sands, gravelly sands		Not needed where sand and gravel filter is used but may be needed with flexible drain tubing and other type filters	
SM (fine)	Silty sands, poorly graded sand-silt mixture			
ML	Inorganic silts and very fine sands, rock flour, silty or clayey fine sands with slight plasticity	Filter needed		
MH	Inorganic silts micaceous or diatomaceous fine sandy or silty soils, elastic silts			
GP	Poorly graded gravels, gravel-sand mixtures, little or no fines		Not needed where sand and gravel filter is used but may be needed with flexible drain tubing and other type filters	With filter—none
SC	Clayey sands, poorly graded sand-clay mixtures	Subject to local on-site determination		Without filter—1.40 ft/sec (0.43 m/s)
GM	Silty gravels, poorly graded gravel-sand silt mixtures			
SM (coarse)	Silty sands, poorly graded sand-silt mixtures			
GC	Clayey gravels, poorly graded gravel-sand-clay mixtures			
CL	Inorganic clays of low to medium plasticity, gravelly clays, sandy clays, silty clays, lean clays			
SP,GP (coarse)	Same as SP and GP		Optional	None—for soils with little or no fines
GW	Well-graded gravels, gravel-sand mixtures, little or no fines	None	May be needed with flexible drain tubing	1.40 ft/sec (0.43 m/s) for soils with appreciable fines
SW	Well-graded sands, gravelly sands, little or no fines			
CH	Inorganic, fat clays			
OL	Organic silts and organic silt-clays of low plasticity			
OH	Organic clays of medium to high plasticity			
Pt	Peat			

Source: U.S. Department of Agriculture, Natural Resource Conservations Service, formerly Soil Conservation Service (1971).

5.3.3 Hydraulic Envelopes. The design of hydraulic envelopes may be less restrictive than the design of filter envelopes. Sand gradations used for concrete as specified by ASTM C33 (fine aggregate) or AASHTO M6 will satisfy the USDA, SCS (1991) hydraulic envelope criteria and will usually satisfy the USDA, SCS (1991) filter envelope requirements for most soils.

5.4 ENVELOPE MATERIALS

5.4.1 Natural Materials

5.4.1.1 Aggregates for drainage envelopes must be selected and sized for maximum permeability and the filter criteria given in 5.3. Aggregates must be chemically and structurally stable and must be free of vegetative matter and bentonitic clay films. Flow rates through aggregates vary greatly depending on gradation.

5.4.1.2 Graded sands may be used as filters if their permeability is sufficient and they meet the filter criteria in 5.3.

5.4.1.3 Slags. Blast furnace and steel mill slags, although not truly natural, belong in this group. Again, sizing and permeability are critical. These materials must be regarded with caution because of the chemical reaction potential of certain of these slags. In some cases, these materials can be cementitious. They can also release materials that can clog drains. Structural stability may also be a concern.

5.4.2 Artificial Materials. Geotextiles used in drainage applications should be nonwoven needle punched, knitted, or spun bonded. Critical parameters are permittivity, AOS (apparent opening size), and survivability during construction. The designer should consider the potential for geotextile clogging, either by small-grain soil particles or due to biological growth. Geotextiles are available and manufactured from polypropylene, polyester, or nylon.

Permittivity values in excess of $1.0\,\text{sec}^{-1}$ should be required. Permittivity, not permeability, should be used because evaluating the quantity of water that would pass through the fabric under a given head over a given cross-sectional area without regard to fabric thickness is important.

The AOS of a fabric is the number of the U.S. standard sieve with openings next larger to the actual size of the geotextile openings as determined by the ASTM D4751 method. The fabric-soil relationship should be $\text{AOS}/D_{85} \leqq 1$.

Two properties critical to the survivability of a geotextile are grab strength and elongation. Tested per ASTM D5034, grab strength should exceed 100 pounds, and grab elongation should not exceed 50%.

CHAPTER 6
HYDRAULICS AND HYDROLOGY

6.1 GENERAL INFORMATION

Successful design of subsurface drainage systems requires an understanding of the behavior of groundwater hydrology and the effects of drainage systems on saturated soils. Furthermore, understanding groundwater collection and discharge properties of drainage systems is essential to ensure desired performance.

6.2 WATER SOURCES

6.2.1 Subsurface Water Sources. In this document, subsurface water is considered to be all water beneath the ground or pavement surface and is sometimes referred to as groundwater.

Soil water is generally of three types: drainable water, plant-available water, and unavailable water. Plant-available water is often referred to as "capillary water," because it is retained by the soil in small soil pores where capillary forces prevent gravity-influenced drainage and is available for plant root absorption.

Drainable water may be considered to be water that readily drains from soil under the influence of gravity. Drainable water moves through soils in direct proportion to the soil's permeability and hydraulic gradient, so low permeabilities result in slow natural drainage of saturated soils.

Unavailable water is held tightly in thin films, surrounding individual soil particles. The strong film bond makes this water nondrainable and unavailable to vegetation. The amount of this hygroscopic water varies with the surface area of the soil particles and, therefore, is highest in clay and organic soils.

The presence of subsurface water can result from direct surface infiltration or water entering the subsoil from adjacent areas. Another potential contributor to soil wetness is artesian water or water moving up through semipermeable soils under piezometric pressure.

Water infiltration in soils is affected by soil type, season of the year, antecedent moisture conditions, type and extent of vegetative cover, surface "crusting" tendency, and characteristics of the particular rainfall event.

6.2.2 Surface Water Sources. Surface water sources that can contribute to groundwater consist primarily of reservoirs, ponds, canals, open drainage ditches, and rainfall. Often the subsurface drainage requirement is a direct result of seepage from one or more of these sources. It is also common for artificial structures to change drainage in ways that can exacerbate subsurface drainage problems. For example, the focused drainage of roof runoff onto flat ground surfaces immediately adjacent to building foundation walls is a common cause of basement flooding. The settlement of disturbed soils adjacent to building foundations over time can exacerbate this condition. Another common problem involves preferential groundwater flow paths created when the pervious bedding and backfill of utility trenches create flow pathways from ponds to areas where this water can cause problems. The designer of subsurface drainage facilities should investigate any probable surface water sources and, if appropriate, incorporate surface drainage improvements into the design.

In certain situations introducing surface runoff into subsurface drains may be necessary, although the practice should be avoided in most cases. When it is necessary, the surface flow inlet must be protected by an effective trash rack and all downstream pipe should be at least 12 inches (30 mm) in diameter to reduce potential for clogging.

6.3 ESTABLISHING THE NEED FOR SUBSURFACE DRAINAGE

6.3.1 General. Excess soil moisture or ponded surface water may be caused by one or more factors. Disregarding runoff from adjacent areas or subsurface aquifers, low soil permeability is the usual cause of extended water retention following precipitation.

Permeability, or hydraulic conductivity, is used as a measure of the soil's ability to transmit water by gravity. Generally, coarse materials, such as sand, are highly permeable and have good transmission rates. Clay soils, however, are usually relatively impermeable, and water retention is long term in the absence of a drainage system. Crude estimates can be made of a soil's permeability by realizing that the passage of water depends greatly on voids in the soil structure, explaining why granular soils with higher void sizes move water better than compact soils with small grain sizes or void sizes.

Soil permeability is best determined by onsite testing. Because of the complex nature of soil composition and its influence on permeability, accurately determining this property without direct measurement is nearly impossible. Whereas laboratory testing procedures (ASTM D2434) have been established, test accuracy is directly influenced by the inherent difficulties related to duplicating the natural undisturbed soil state. Several reliable in situ testing methods have been developed for both saturated soils and unsaturated soils as described in the U.S. Bureau of Reclamation's *Drainage Manual* (USBR 1993). At least three in situ tests should be conducted in each major soil group that will influence the drainage system. More tests are recommended for very large systems.

It is often necessary to involve groundwater and geology professionals in the proper design and evaluation of subsurface drainage systems.

6.3.2 Removal Criteria for Different Environments and Climates. Subsurface drainage, when viewed as water management, is used agriculturally as a method to improve the soil environment for plants and to eliminate operational hazards and nuisances. Benefits in economy and safety are realized.

The rate at which surplus groundwater must be removed relates primarily to the moisture and air requirements of vegetation.

Drainage of urban areas generally relates to lawns, parks, and recreational turf and usually involves faster removal rates than for agriculture. Faster water removal from the surface and topsoil zone allows quicker access to turf surface and minimizes athletic disturbance of vegetation. Aeration of vegetation is essential to its health and durability, and some areas require enhanced drainage for salinity control. Groundwater control is necessary for the correct performance of septic tank absorption fields and for detention ponds.

Climatic conditions must be considered. Soils in humid regions often require more extensive drainage systems than soils in arid regions. Temperature and humidity conditions interact with soil characteristics to influence moisture control requirements.

Comprehensive data for the selection of optimum drawdown rates for nonagricultural drainage systems are not yet available. Local turf specialists should be of assistance, and regional data should identify distinctive climatic conditions such as humidity and evapotranspiration rates.

6.3.3 Special Requirements for Paved Surfaces. Vehicular traffic on pavements with a saturated subbase results in early deterioration. Water enters through openings in the pavement from high groundwater conditions or as seepage from adjacent regions. Correctly designed paved areas should have highly permeable base or subbase construction and drainage systems to promote rapid outflow of infiltrated water.

6.4 BASIC SUBSURFACE DRAINAGE THEORY

6.4.1 Controlled Water Table Elevation. Subsurface drainage is accomplished by placing an artificial channel below the water table so that the hydraulic head of the channel is less than that of the soil to be drained. The hydraulic head differential creates a hydraulic gradient in the direction of the artificial channel, depressing the phreatic line (also called free water surface; see Fig. 6-1) in the vicinity of the artificial channel. The constant removal of water flowing into the drainage sink maintains the hydraulic head differential, thus maintaining the depressed phreatic line.

The hydraulic gradient and the hydraulic conductivity of the soils to be drained govern the rate at which water moves toward the sink. Control of water is accomplished by controlling the hydraulic gradient. Therefore, flow is regulated by adjusting the depth of the sink and the spacing between sinks and by locating the sinks to take advantage of the more permeable soils in the area to be drained.

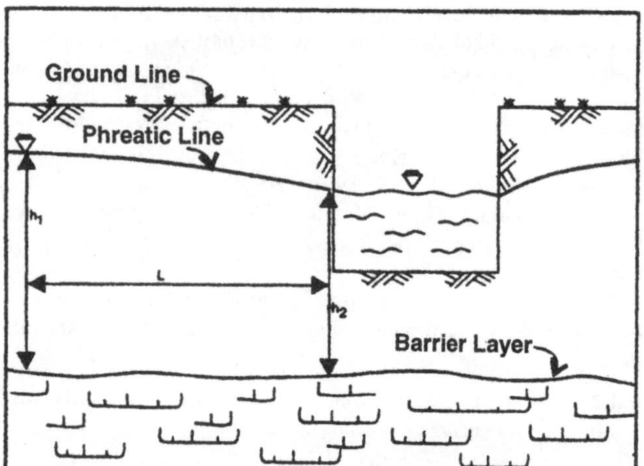

FIGURE 6-1. Soil water movement

6.4.2 Drainable Versus Nondrainable Groundwater. Drainable water, or "gravity water," is that which is free to move through the soil by the force of gravity or hydraulic gradients. When all voids are filled with water, the soil is considered to be saturated. Some water is held in the soil against gravity. It includes the film of water left around the soil grains and the water filling the smaller pores after gravity water has drained off. This water is considered to be nondrainable groundwater.

6.4.3 Drainage Formulas and Recommended Practices. Selection of the proper drainage system to control the anticipated flow will vary with the specific application, but in all applications the procedure begins with a determination of the allowable subsurface water elevation and the drain depth. Spacing of drains can then be calculated, and a discharge can be computed for each drain.

Most methods for estimating drain spacing are empirical and were developed to meet specific characteristics of a particular area. Some methods are based on assumptions of steady-state flow conditions where the hydraulic head does not vary with time. Other methods assume transient flow conditions where the hydraulic head changes with time.

It is occasionally necessary to perform indepth analysis of the three-dimensional groundwater problem to come up with a cost-effective solution. Several numerical model techniques are available for these situations.

6.4.3.1 Steady State. The most widely used steady-state formula for subsurface drainage spacing is the ellipse equation. In the ellipse equation, the depth to the drain is first established, then spacing of the drains is computed by the formula. The ellipse equation is based on an assumption that the streamlines of flow in a gravity drainage system are horizontal, and the velocity of flow is proportional to the hydraulic gradient or the free water surface. Although approximate, these assumptions may closely approach actual conditions in certain sites. For this reason, use of the ellipse equation should be limited to the following conditions:

- Where recharge is primarily from the soil surface.
- Where the groundwater flow is known to be largely in a horizontal direction, such as stratified soils with relatively permeable layers acting as horizontal aquifers.
- Where a barrier layer underlies soil and subsoil materials at relatively shallow depths (less than twice the depth to the drain), which restricts vertical flow and forces the groundwater to flow horizontally toward the drain.
- Where open ditches are used or where drains with sand and gravel filters or porous trench backfill are used.

These conditions cause a minimum of restriction to flow into the drains and minimize convergence of flow at the drains.

The ellipse equation (ASCE 1998) is usually expressed in the form:

$$S = \sqrt{\frac{4K_s\left(m^2 - a^2\right)}{q}} \qquad (6\text{-}1)$$

where

S = drain spacing, ft (m);
K_s = average hydraulic conductivity, in./hr (m/day);

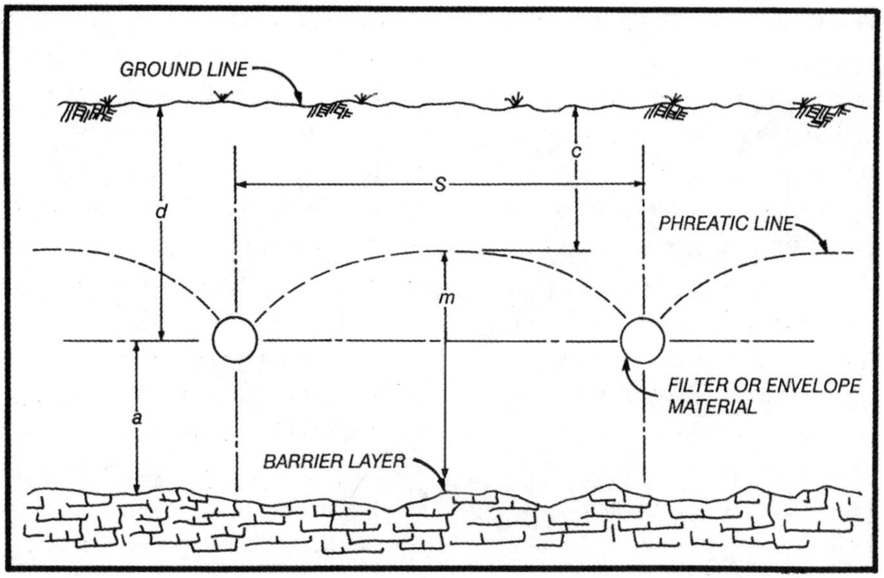

FIGURE 6-2. Cross-sectional view showing symbols used in ellipse equation

m = vertical distance, after drawdown, of water table above the barrier at midpoint between drain lines, ft (m);
a = depth of barrier layer below drain, ft (m); and
q = drainage coefficient (steady-state surface recharge rate), in./hr (m/day).

Details of the symbols and parameters for Eq. 6-1 are shown in Fig. 6-2. A correction for convergence losses should be used. A convergence correction developed by Hooghoudt (USBR 1993) considers the loss in head required to overcome convergence in the primary spacing calculation. The calculations derive an equivalent depth to barrier, which is then inserted into the ellipse equation in place of the actual depth to barrier, and a new spacing is calculated. The procedure for computing the adjustment for convergence is discussed in U.S. Bureau of Reclamation and U.S. Department of Agriculture references (see Chapter 3) with examples presented for use of both steady-state computations and transient-state computations.

6.4.3.2 Transient State. Transient drainage formulas have been developed, the most notable of which are those developed by Ernst, Hooghoudt, Dumm, and Van Schilfgaarde. These drainage models are considerably more complicated and apply to more complex types of geologic and hydrologic conditions. The transient-state drain spacing processes are more detailed in that they require the designer to develop a pattern of recharge events to simulate anticipated future conditions. In return, the processes facilitate analyses of the drain design for economic assessment and risk assessment. Economic assessment allows the designer to select the most economical depth and spacing, while the risk assessment reveals the probability that the water table will rise above design limits with any given design. Transient-state spacing computations should be corrected for convergence using the same methods described for steady-state computations.

The particular formula or model selected for computing the spacing of relief drains is influenced by site conditions and experience from drains in similar soils and climates. Reference is made to background material presented by TeKrony, Sanders, and Cummins (2004).

6.5 SUBSURFACE DRAIN APPLICATIONS

6.5.1 Soil Consolidation Using Wick Drains. Compressible soils with high water content are generally unsuitable on construction sites selected for any structure requiring a stable foundation. Because the damaging effects of soil settlement are well documented, the practice of soil stabilization through the process of consolidation has been widely accepted. This process fundamentally involves the development of a surcharge loading on the site to produce excess pore pressures in the water, forcing water from the soil pores. According to the theory of consolidation, the results vary inversely with the soil permeability and directly to the square of the longest drainage path. In most cases, especially where fine silts and clays are present, the process is extremely long if unaided by a drainage system.

Establishing vertical trenches and filling them with sand or inserting geocomposite wicks into the soil to be consolidated shortens the drainage path and accelerates the rate of consolidation. The design of sand drain systems is commonly being replaced by smaller geocomposite wicks. This permits quicker and more reliable construction.

6.5.2 Groundwater Removal with Relief Drains. Relief drains are broadly considered to be any product or construction that accelerates the removal of drainable subsurface water. Such drains are commonly required in soils whose properties result in extended saturation periods following rainfall events, thus affecting infiltration and runoff, vegetative health, and seepage forces on underground structures. In the broadest sense, relief drains can take the form of trenches or paved channels, but they are more commonly a buried product or structure that is systematically designed and located to serve as a subsurface water collector or serve as both a surface water collector and a subsurface water collector. Relief drains also discharge the collected water to a selected area or structure.

Relief drain design and construction has evolved over many centuries to include sand- and aggregate-filled trenches (both with and without perforated pipe in the lower portion of the select fill), pipe with small perforations through the wall surfaces (installed either with or without a geotextile covering),

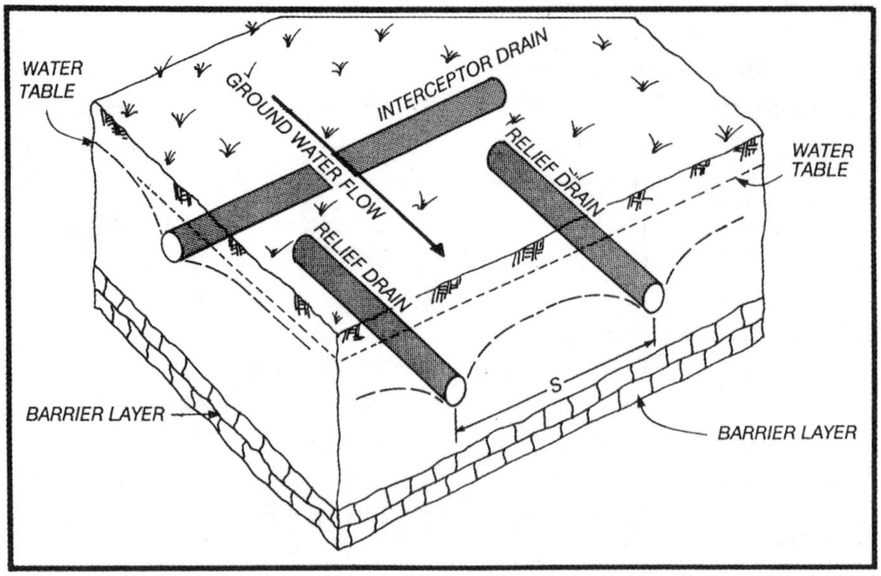

FIGURE 6-3. Isometric profiles of relief and interceptor drains

solid-wall pipe installed with gaps between each length (called "open joint pipe"), and geocomposite fin drains (formed plastic interior members overwrapped with filter textiles). Figure 6-3 shows one configuration of these products and systems that provide groundwater removal from turf areas and protect subsurface structures from seepage forces.

Although specifications for open-joint pipe are included in 1.1, 8.1.1, and 8.1.4, the use of such pipe is not recommended. Entrance of fines through the joints and sediment deposition within the pipe can be problematic. And because the inflow is not uniform along the length of the pipe, open-joint pipe is not readily amenable to hydraulic analysis.

Drainage systems must serve two primary functions over long periods of time: Collect groundwater and then discharge the collected water to a selected point. In nearly all cases, gravitational force and the resulting driving energies are relatively small, requiring care in design and placement to avoid major performance reduction.

6.5.3 Seepage Control with Interceptor Drains. Seepage is the slow movement of gravitational water through soil and rock; the flowing water creates a frictional drag force proportional to flow velocity. Seepage forces have been significant contributors to the failures of dams and other earthen structures, in addition to subsurface structures such as foundations and retaining walls.

In broad terms, there are three ways to deal with seepage groundwater. In a few cases, the structure can be designed to allow passage of groundwater without endangering the integrity and performance of the structure. The remaining two seepage control methods either keep the water out or use drainage methods to control water removal. The techniques for shielding a structure involve wall or curtain construction.

Controlling seepage via drainage techniques essentially involves the placement of a subsurface drain so that it intercepts the groundwater and protects the structure, which otherwise would be affected by the seepage force. As shown in Fig. 6-3, the general design of such interceptor drains is normally perpendicular to the slope of the water table. A good example of the use and effectiveness of interceptor drains is the long-standing practice of surrounding building foundations with highly permeable materials such as sand or aggregate. Retaining wall designs also frequently have nearly full-height drainage material to intercept seepage water. A suitable outlet must be provided for the collected water.

There are many design similarities between relief drains and interceptor drains. Likewise, the precautions regarding long-term performance are similar. In both cases, extreme care is required to balance discharge capacity with inflow expectations and to evaluate the soil environment and inflow area materials to protect against reductions of inflow or discharge capacities.

Also, similar to subsurface relief drains, geocomposites of sufficient size and geometric shape may be employed as interceptors, and these are particularly adaptable to rigid flat surfaces such as retaining walls and foundations.

6.5.4 Water Control of Pavement Structural Sections. Concrete and asphalt areas (including highways, streets, parking lots, and parks) usually are supported by an underlay of sand or stone, which should provide both structural stability and drainage for the total structure. However, significant amounts of water can enter this base or subbase course from a variety of sources and, without the opportunity to drain, remain entrapped for extended time periods. Combinations of vehicular loadings and freeze-thaw cycles on pavements with saturated base layers cause large hydraulic and ice pressure forces that accelerate breakdown of the pavement. Such deterioration is widely experienced on heavily traveled roadways and is considered the major cause of deterioration on most highway systems in the United States. Other paved areas also suffer shortened lives due to inadequate subgrade drainage. Included in this group are airport runways and taxiways, residential streets, parking lots, sidewalks, and tennis courts. Any outdoor pavement should be designed for rapid removal of infiltrated water in the base course following rainfall. Federal Highway Administration (FHWA 1992) training materials suggest that the saturation level be reduced to 85% or less within two hours after rainfall ceases.

The design of pavement drainage generally involves both the base layer and a specific drainage product or system. Designs require a drainable structural base or subbase course, employing open-graded materials. Low-permeability sand bases, except for small areas with light loadings (nonvehicular), may be inadequate.

Open-graded permeable structural courses slope toward the drainage system, where water from the course is collected and discharged. Conventional roadway design is to slope the course following the roadway center crown or elevation and place a collector drain system at the downslope edge, which may be both sides of a relatively flat roadway or one side of a superelevated curve. Once the aggregate layer drains freely into the edge drain(s), the drain must function as a relief drain of significantly greater length. Because of the long runs in typical edge drain installation, side outletting into a drainage ditch or storm drain is necessary for outflow-inflow continuity. Outlet spacing is typically several hundred feet (multiples of 30 m) on centers for pipe or high-column edge drain systems on major highways. Catch basins provide a convenient opportunity to direct flow from the subsurface edge drain into the storm drain system.

6.6 INFLOW-OUTFLOW CONTINUITY

For adequate subsurface drainage system performance in any application, designs balancing outflow and inflow are essential. All sources of inflow must be identified and quantified. Inflow sources are most commonly these: (1) surface infiltration (rainfall, snowfall, and irrigation sources); (2) water transfer from adjacent areas (springs, waterfalls, and watersheds); and (3) high water table.

The flow of water through soil and/or pervious base material into subsurface collectors, accompanied by discharge and disposal, depends on several factors: (1) soil permeability (hydraulic conductivity); (2) flow energy losses at collectors (such as siltation of filter envelopes or fabric, convergence of flow lines, and entrance resistance of perforations); and (3) hydraulic characteristics of the subsurface drain system.

Soil permeability is carefully evaluated, because it is a requisite element to subsurface drainage performance. Flow reduction potential due to convergence and siltation must be considered because these also contribute significantly to the result. Equally important is the determination of hydraulic properties for the collector and drain systems to permit unrestricted discharge of collected water.

6.7 INFLOW TO COLLECTORS

The following methods should be used for drain sizing.

6.7.1 Relief Drains. Procedures for calculating the flows to relief drains involve prior computed drain spacings. The spacing is computed using Eq. 6-1 as described in 6.4. The area served by parallel relief drains is equal to the spacing times the length of the drain plus one-half the spacing. The total discharge can be expressed by the following steady-state equation (USDA 1969; USDA, SCS 1971; ASCE 1998) (customary units):

$$Q_r = \frac{qS(L + S/2)}{43{,}200} \tag{6-2}$$

where

Q_r = relief drain discharge, cubic feet per second;
q = drainage coefficient, inches per hour;
S = drain spacing, feet; and
L = drain length, feet

or in SI units:

$$Q_r = qS(L + S/2) \tag{6-3}$$

where

Q_r = relief drain discharge, cubic meters per day;
q = drainage coefficient, meters per day;
S = drain spacing, meters; and
L = drain length, meters.

Figure 6-4 shows a parallel relief drain system; the shaded area indicates the area served by one of the relief drains. The value of Q_r is for the drain in its entirety (i.e., at its discharge point to a collector drain). The flow into the relief drain will be uniform along its length, with the discharge at any point along the drain proportional to the length. This suggests that collector drains receiving flow from several relief drains can be sized according to the flow carried at the various points along the collector. Depending on the size, shape, and configuration of the area to be drained, this variable sizing could result in significant cost savings.

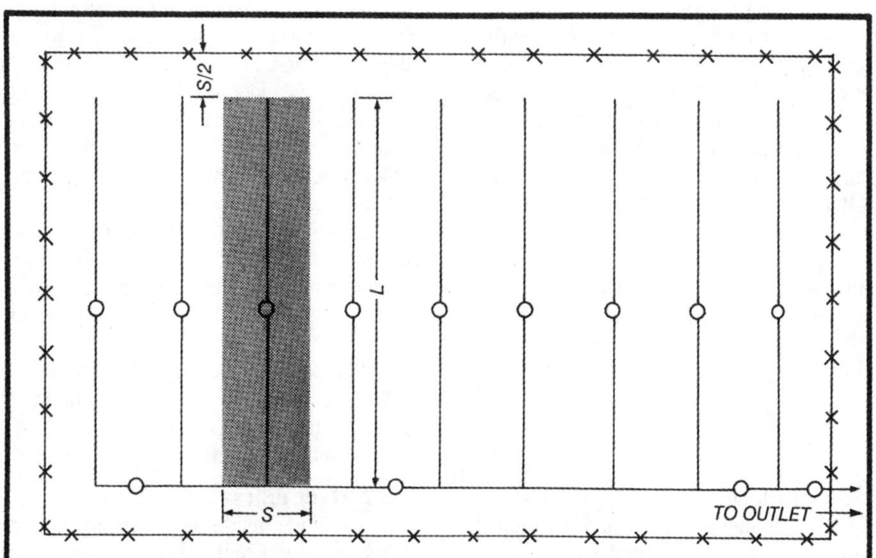

FIGURE 6-4. Plan view of relief drain system

The inflow to drains may also be calculated using transient-state methods. The transient-state equations require a known depth to groundwater, a known depth to barrier, and a known hydraulic conductivity, all of which can be directly measured. The drainage coefficient, which cannot be measured directly, is not required. The transient-state equations may be used to verify the estimated drainage coefficient by comparing the results from the two methods, steady-state and transient-state, thereby lending credence to the drain spacing.

When a single drain line is used rather than a system of spaced drains, regional groundwater flow may account for a large part of the flow to the drain, and the contributing area is usually not known. In these cases, Eqs. 6-4 or 6-5, which do not require a known contributing area, can be used to determine the inflow rate to the drain.

The transient state equations are (USBR 1993)

$$q_p = \frac{2\pi K y_0 D}{86{,}400L} \quad \text{(Eq. 6-4, for drains above the barrier)}$$

and

$$q_p = \frac{4KH^2}{86{,}400L} \quad \text{(Eq. 6-5, for drains on the barrier)}$$

where

q_p = discharge from two sides per unit length, in cubic feet (cubic meters) per second per foot (meter);
y_0 or H = maximum height of water table above the drain invert in feet (meters);
K = weighted average hydraulic conductivity of the soil profile between the water table and the barrier in feet (meters) per day;
D = average flow depth ($D = d + y_0/2$) in feet (meters);
d = the distance from drain to barrier in feet (meters); and
L = drain spacing in feet (meters).

Reference is made to background material presented by TeKrony, Sanders, and Cummins (2004).

6.7.2 Wick Drains and Chimney Drains. Groundwater moves into the drain from the surrounding soils due to higher pore pressures; soil consolidation rates are accelerated by the escape of water from the soil voids. Volume flow rate to these drains is a function of the number and type of alternating soils in the profile; the permeability and porosity of the soil; extent of soil disturbance during installation of the drains; the radial flow resistance of the drain; blinding/clogging potential of the sand or geotextile wrap; in the case of geocomposite drains, the amount of free (open) inflow surface; the effective size of the drain; and the number of drains installed. The conjunctive hydrologic and hydraulic design of wick drains is discussed in 6.8.7.

6.7.3 Interceptor Drains. In calculating the flow to an interceptor drain from upslope sources, the Darcy equation may be used. The flow per unit length into an interceptor drain may be calculated using the following equation (USBR 1993):

$$q_u = K_c IA[y/(y+d)] \qquad (6\text{-}6)$$

where

q_u = volume rate of flow into drain per unit length, cubic feet per second per foot (cubic meters per second per meter);
K_c = weighted average unit hydraulic conductivity of the strata above the barrier, feet per second (meters per second);
I = slope normal to groundwater contours, feet per foot (meter per meter);
A = saturated area in square feet per foot (square meters per meter) of flow in a unit length of drain;
y = height of maximum water surface immediately above proposed drain, feet (meters); and
d = distance from drain invert to impermeable barrier, feet (meters).

K_c is computed from:

$$K_c = \frac{K_1T_1 + K_2T_2 + \ldots 1K_nT_n}{T_1 + T_2 + \ldots 1T_n} \qquad (6\text{-}7)$$

where

K_i = hydraulic unit conductivity for layers K_1 through K_n of the soil profile, feet per second (meters per second); and
T_i = thickness of layers T_1 through T_n, feet (meters).

Generally, the maximum water table height would be used to obtain the saturated depth from which K_c is obtained. This same depth would be used to obtain the area A for a unit width. The plane along which the area must be obtained is parallel to the contours or perpendicular to the direction of flow.

6.7.4 Perimeter Drains and Edge Drains. Perimeter drains and edge drains function similarly to relief drains. Used longitudinally on roads and airport runways, edge drains are designed to quickly remove excess water from the base or subbase course and subgrades of pavement structures to prevent premature deterioration of the pavement.

Flow into an edge drain or a perimeter drain is similar to that of the interceptor drains discussed in 6.7.3, except most flow enters from the pavement side of the drain. Side outlets for edge drains must be spaced according to the estimated flow in the edge drain and the hydraulic flow capacity of the drain. In using some of the newer geocomposite fin drains, laboratory tests for inflow and capacity of each specific product being evaluated must be used for comparisons of appropriateness because specific criteria have not been adopted by consensus.

6.8 HYDRAULICS OF SUBSURFACE DRAINS

6.8.1 Outflow of Collected Water. The design of a subsurface drainage system includes the hydraulic design of all conduits that collect the excess water from the soil and deliver the water to the outlet for safe discharge. The hydraulic capacity of the drains should be checked at all points of size modification and at lateral connections. Subsurface drains are normally designed as open channels with a free water surface (at atmospheric pressure) inside the pipes; this incorporates the implicit assumption of decoupling of the groundwater hydrology from the drain conduit hydraulics and enables a rational, quantitative design. The purposeful or inadvertent design of the drain conduit for pressurized flow can result in nonuniform inflow at one end of the spectrum and uneconomical design at the other (Graber 2007a). The amount of water collected depends on the drainage coefficient for the area and the size of the area contributing water to the drain. If lateral seepage is occurring from outside the area drained, the conduits must be large enough to carry the additional water that will enter the system.

6.8.2 Hydraulics of Drainpipes. The specification of a free water surface inside the drainpipe as a design factor means that all water moves in the system in response to gravity and frictional forces. The water in the saturated soil above the drain moves by gravity toward the water level inside the drain, developing a classic drawdown curve and a water table that slopes toward the drain. Once inside the drain, the water moves down gradient in the pipe according to open channel hydraulic principles. Under free discharge conditions, there is a progressive drop in the energy grade line in the direction of the outlets. For steady uniform flow, the water surface and the energy lines are parallel to the bottom slope of the pipe. However, in a subsurface drain, the flow rate increases in a downstream direction so the flow rate to be carried by the pipe at any point along its length is not a constant and the slope of the energy line also varies along the length of the pipe. Drainpipes are usually installed on a uniform grade. The maximum carrying capacity for the drain is calculated at the end of each section, that point at which maximum flow occurs in a pipe section having a uniform pipe size and a constant bottom slope. Consider energy losses due to friction, changes in flow direction, changes in the size or shape of the pipe, control or measurement devices, inlet and outlet head losses, and momentum transfer. Momentum transfer required to accelerate water entering the pipe at nonzero angles to the flow direction in the pipe must also be considered at junctions.

An adequate pipe bottom slope and adequate minimum flow velocity (but see the following) are important in subsurface drains to minimize sediment deposition. Slopes of at least 0.001 are usually adequate. Excessively steep slopes may cause erosion of the soil outside the pipe perforations and result in failure of the drain. Some drains, such as foundation drains, may have to be laid with a zero slope. However, a hydraulic grade line will develop to remove the water as long as an adequate outlet is available and the pipe is properly designed. Information regarding drain grades and velocities and related aspects of drain filters or envelopes (which can eliminate the need for minimum velocities) can be found in Section 16 of USDA, SCS (1971) at pages 4-81 to 4-83, including Table 4-6 and pages 4-90 to 4-96,which includes a table identical to Table 5-1.

The potential inadequacies of basing the design of conduits with spatially increasing flow on the assumption of uniform flow (i.e., Manning's equation alone) are addressed by Graber (2004). A spatially varied flow calculation using numerical methods may be made using the ordinary differential equations for spatially varied flow (Chow 1959) to ensure that the hydraulic grade line remains inside the conduit and that the design is economical. In the latter case, the Manning equation is used to calculate a friction slope at each computational section along the conduit. Figure 6-5, based on Graber (2004), provides a chart that can be used for design of circular conduits of single pipe diameter with uniform inflow and free discharge conditions (tailwater elevation less than downstream critical depth elevation). The variables in Figure 6-5 are as follows:

n = Manning roughness coefficient;
d_o or D_o = pipe diameter (ft, m);
S_o = pipe slope;
L = pipe length (ft, m);
Q_L = rate of flow (cfs, m^3/sec);
g = gravitational acceleration (32.2 ft/sec^2, 9.81 m/sec^2); and
K_n = 1.486 for customary units, 1 for SI units.

Figure 6-5 is based on a ratio of maximum flow depth y_{max} to pipe diameter of $y_{max}/d_o = 0.95$.

For uniform spatial inflow, constant pipe diameter, turbulent flow, and free discharge, Graber (2004, 2007b) shows that the uniform-flow approximation is conservative for:

$$S_o L/d_o > 5,\ y_{max}/d_o = 0.95,\ \mathbf{F}_L = 1$$

in which $\mathbf{F}_L$ = Froude number at downstream end. $\mathbf{F}_L = 1$ requires that the tailwater be at or below the critical-depth elevation at the downstream end (free discharge). Additional perspective

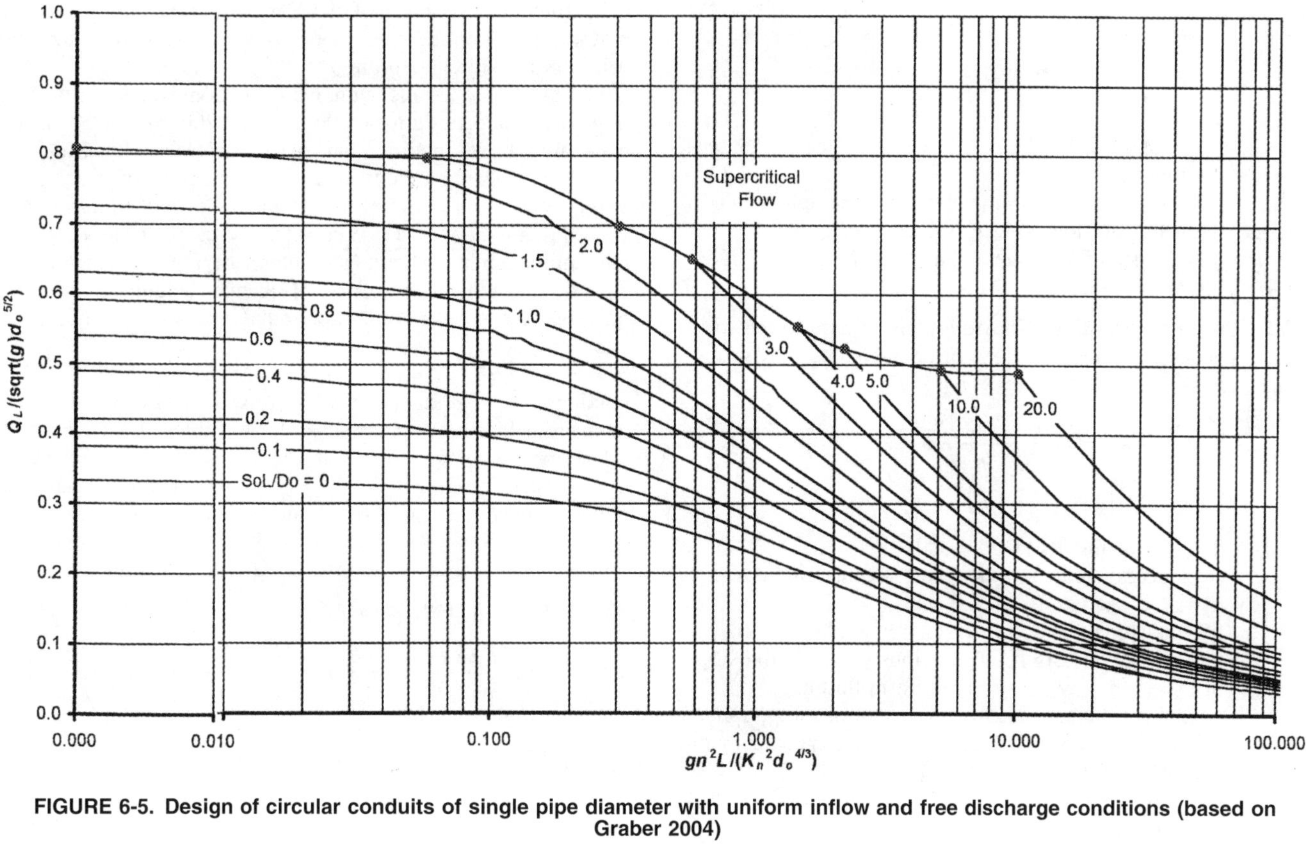

FIGURE 6-5. Design of circular conduits of single pipe diameter with uniform inflow and free discharge conditions (based on Graber 2004)

Table 6-1. Recommended Design Values of Manning Roughness Coefficients for Closed Conduits and Open Channels (ASCE 2007)

Conduit Material	Manning's *n*
Asbestos-cement pipe	0.011–0.015
Brick	0.013–0.017
Cast-iron pipe	
Cement-lined and seal-coated	0.011–0.015
Concrete (monolithic)	
Smooth forms	0.012–0.014
Rough forms	0.015–0.017
Concrete pipe	0.011–0.015
Corrugated-metal pipe	
Plain annular	0.022–0.027
Plain helical	0.011–0.023
Paved invert	0.018–0.022
Spun asphalt lined	0.011–0.015
Spiral rib metal pipe (smooth)	0.012–0.015
Ductile iron pipe (cement lined)	0.011–0.014
Plastic pipe (corrugated)	
3–8 in. (75–200 mm) diameter	0.014–0.016
10–12 in. (250–300 mm) diameter	0.016–0.018
Larger than 12 in. (300 mm) diameter	0.019–0.021
Plastic pipe (smooth interior)	0.010–0.013
Vitrified clay	
Pipes	0.011–0.015
Liner plates	0.013–0.017

regarding this approximation and its relation to other references is given in Graber (2007b).

For more complex systems, involving, for example, changes in pipe diameter or downstream submergence, a computer program may be used for design and analysis purposes (Graber 2004).

6.8.3 Manning Equation. The following equation was proposed by Robert Manning in 1889 for open-channel flows (customary units):

$$Q = \frac{A\,1.49\,R^{2/3}S^{1/2}}{n} \tag{6-8}$$

where

Q = flow rate, cubic feet per second;
A = cross-sectional area of the water stream, square feet;
R = hydraulic radius: $R = A/WP$, feet;
S = friction slope or slope of the energy grade line, feet per foot (which, as discussed previously, equals the slope of the drain only for steady uniform flow);
WP = wetted perimeter, which is the length of the periphery of the cross-sectional shape of the liquid in contact with the pipe wall surface, feet; and
n = Manning roughness coefficient for the drain product (refer to Table 6-1).

Or in SI units:

$$Q = \frac{AR^{2/3}S^{1/2}}{n} \tag{6-9}$$

where

Q = flow rate, cubic meters per second;
A = cross-sectional area, square meters;
R = hydraulic radius, meters;
S = friction slope, meters per meter; and
n = roughness coefficient, as given in Table 6-1.

6.8.4 Hydraulics of Aggregate Drains. In all but extremely coarse aggregates, flow velocity is low and laminar, thus permitting use of the basic Darcy equation:

$$Q = KiA \tag{6-10}$$

where

Q = flow rate, cubic feet per second (cubic meters per second);
K = hydraulic conductivity, feet per second (meters per second);
i = hydraulic gradient (approximately equal to the slope [Δh/l)]), feet per foot (meter per meter); and
A = cross-sectional area of the aggregate-filled trench, square feet (square meters).

The most significant variable is conductivity, which principally depends on the voids established by the aggregate's particle size and gradation. Aggregate conductivity can vary widely. Laboratory testing to confirm conductivity of selected aggregate versus the design requirements is highly recommended.

The amount of fines in the aggregate mix affects conductivity. Generally, higher fines content results in lower conductivity, indicating the desirability of clean, washed stone. Safeguards are required to minimize passage of excessive fines from adjacent soil into the aggregate drain.

For aggregate drains with Darcian, homogeneous in-plane flow, a rational design method (Graber 2007b) can be developed by slight modification of a derivation by McEnroe (1993). For Darcian flow in a sloped medium of rectangular cross section and constant width, the following dimensionless variables are defined in terms of S_o = channel slope; y_{max} = maximum depth of flow; y_L = depth of flow at downstream end; L = horizontal projection of channel length; r = uniform rate of inflow per unit horizontal area; K = hydraulic conductivity of drain medium: $Y'_{max} = y_{max}/L$; $Y'_L = y_L/L$; $R' = (1+S_o^2)r/K$; $A' = (S_o^2 - 4R')^{1/2}$; and $B' = (4R' - S_o^2)^{1/2}$. The nondimensional primed variables are so denoted to distinguish them from McEnroe's (1993) similar variables. A free-drainage Numerov boundary condition (McEnroe 1993, Harr 1962) is assumed, for which $Y'_L = R'$; this requires that the tailwater elevation not exceed an elevation corresponding to $y_L = rL(1+S_o^2)/K$. The solution utilizing these variables, presented in the following, admits the case of zero slope:

$$Y'_{max} = (R' - S_oR' + R'^2)^{1/2} \bullet \exp\left[\frac{S_o}{B'}\tan^{-1}\left(\frac{2R' - S_o}{B'}\right) - \frac{S_o}{B'}\tan^{-1}\left(\frac{2R'/S_o - S_o}{B'}\right)\right], \quad S_o^2 < 4R' \tag{6-11a}$$

$$Y'_{max} = \frac{R'}{S_o}\frac{S_o^2 - 2R'S_o}{S_o^2 - 2R'}\exp\left[\frac{2R'S_o^2(S_o - 1)}{(S_o^2 - 2R'S_o)(S_o^2 - 2R')}\right], \quad S_o^2 = 4R' \tag{6-11b}$$

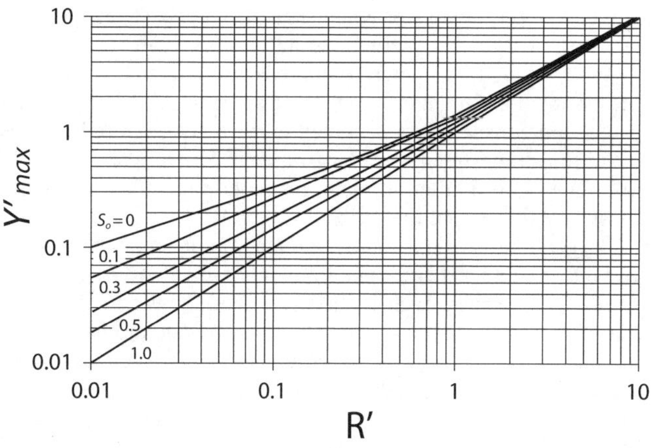

FIGURE 6-6. Rectangular aggregate drain chart with uniform inflow and free discharge (Graber 2007b)

$$Y'_{\max} = (R' - S_o R' + R'^2)^{1/2} \left[\frac{(S_o - A' - 2R'/S_o)}{(S_o + A' - 2R')} \frac{(S_o + A' - 2R'/S_o)}{(S_o - A' - 2R')} \right]^{S_o/(2A')}, \quad S_o^2 > 4R' \tag{6-11c}$$

A plot of these relationships is presented in Fig. 6-6. The maximum depth occurs at the upstream end for $S_o = 0$, moving downstream as S_o increases until at $S_o = 1$ the maximum depth is at the downstream end and $Y'_{\max} = Y_L = R'$ is in conformance with Fig. 6-6. Values of $S_o > 1$ will still have $Y'_{\max} = Y_L = R'$. As suggested by Fig. 6-6, values of $R' > 10$ (well within the range of practicality for aggregate drains) will have $Y'_{\max} \cong Y_L = R'$.

Denoting the measured full-flow transmissivity by T_f, the transmissivity is related to conductivity by $T_f = Kb$. Thus the test values of T_f and known b give $K = T_f/b$, which can be used in the calculation of R' for use with Eqs. 6-11 and Fig. 6-6. The factor of safety should be selected based on the accuracy to which the hydrology (flow rate) is known, the practical importance of inaccuracies in assumed hydrology, and the accuracy to which drain-medium conductivity is known. If a permeameter or other full-flow test device of cross-sectional dimensions smaller than the full-scale cross section is used to estimate conductivity, the conductivity will be higher in the permeameter; that effect can be minimized by using a permeameter with a sufficiently high ratio of permeameter diameter to medium particle diameter or a sufficiently high Darcy number = square of permeameter hydraulic diameter divided by b (Somerton and Wood 1988). Based on Somerton and Wood (1988), a ratio of 100:1 permeameter diameter to the largest particle diameter should give a conductivity within 1% of that for an infinite porous medium. A Darcy number > 10^5 provides an alternative criterion. For a permeameter with a circular cross section, using the least stringent of these criteria would be reasonable. Otherwise a suitable factor of safety should be used for conductivity determined by permeameter or otherwise. For design, factors of safety are discussed by Koerner (2005) in the context of flow-rate reduction factors.

Departure from Darcian flow occurs as an appropriately defined Reynolds number increases, first because inertial forces become significant while the flow remains laminar and, with still higher Reynolds numbers, as the flow becomes turbulent in an increasing portion of the porous medium (Davis and DeWiest 1966, Ward 1964). Ward (1964) expressed the initial departure from Darcian flow in terms of a Reynolds number $\mathbf{R}_k = Vk^{1/2}/\nu$ in which V = macroscopic velocity (also referred to as Darcy velocity or specific discharge), k = permeability, and ν = kinematic viscosity. Permeability is related to conductivity by $k = \nu K/g$, in which g = gravitational acceleration. Because $V = rL/y$ with y = flow depth, and using other terms previously defined, the Reynolds number can be expressed as

$$\mathbf{R}_k = \frac{R'K^{3/2}}{(y/L)(1+S_o^2)(\nu g)^{1/2}} \tag{6-12}$$

The Reynolds number will be a maximum at the minimum flow depth, for which $y/L = y_L/L = Y_L = R'$, so the Reynolds number can be conservatively expressed as

$$\mathbf{R}_k = \frac{K^{3/2}}{(1+S_o^2)(\nu g)^{1/2}} \tag{6-13}$$

Ward (1964) suggested that to limit the error in hydraulic gradient to 10%, $\mathbf{R}_k$ should not exceed 0.182. Ward's criterion, with $\nu \cong 10^{-2}\,\text{cm}^2/\text{s}$ and $g = 9.81\,\text{m/s}^2$ in Eq. 6-13 gives maximum $K = 0.7(1+S_o^2)$ cm/s = O(1) cm/s, corresponding to the upper end of the range for clean sand and the lower end of the range for gravel (Davis and DeWiest 1966). For typical values of r, that will correspond to R' greater than approximately 0.01, which is the lower range of R' plotted on Fig. 6-6. The value of $\mathbf{R}_k$ should be checked for each specific design.

For the transitional and turbulent regimes, Ward (1964) gave relationships that can be expressed in terms of variables used here as follows:

$$i = f_k \frac{V^2}{\sqrt{g\nu K}} \tag{6-14}$$

$$f_k = \frac{1}{\mathbf{R}_k} + C_1 \tag{6-15}$$

in which f_k is the Fanning friction factor for porous media and C_1 is the fully-turbulent friction factor given as $C_1 = 0.550$ by Ward. Rumer and Drinker (1966) suggested on theoretical grounds that C_1 is not a universal constant as Ward indicated but rather that it will vary slightly for different media. Venkataraman and Rao (1998) gave values of C_1 varying by a factor of more than 10 for different media. They suggested that relating C_1 to media properties needs further study, although estimates can be made in some cases from their tabulation of data together with original sources that they reference and others. Equations 6-14 and 6-15 can be used in conjunction with appropriate test data. Friction factor decreases with increasing Reynolds number per Eqs. 6-14 and 6-15; therefore a conservative and practical approach for aggregate drains is to use a test Reynolds number that is less than the Reynolds number at the maximum-depth location under design conditions.

Solutions for geotextile drains or laminar-flow aggregate drains with tailwater higher than that corresponding to the Numerov condition, or for aggregate drains with transitional and turbulent flow, can be obtained for known flow distribution and downstream flow depth using the numerical computational procedures described under "Numerical Formulation" of Graber (2004). In doing so, the friction slope S_f should be taken as i from Eq. 6-14, and the velocities should be taken as the mean interstitial velocity given by $Q/(Ap)$, with Q = flow rate, A = wetted cross-sectional area, and p = medium porosity.

Additional perspective regarding these methods and their relation to other references is given in Graber (2007b).

6.8.5 Hydraulics of Geotextile Drains. Fluid flows through geotextiles serving as a filter, with the primary function of the fabric being to restrain solids. The measure of a fabric's ability to permit water flow perpendicular to its plane is termed "permittivity."

Nonwoven geotextiles are capable of transmitting fluids within the plane of the fabric. Care should be taken to consider the potential for fabric clogging due to small soil or other particles in the system, or due to biological growth. Nonwoven resin-bonded and nonwoven needle-punched geotextiles are especially suitable for drainage applications (Koerner 2005). Planar flow depends on the nature of the fabric, the gradient (slope), and the compressive forces on the material. Planar flow properties for any geosynthetic may be determined from ASTM D4716 with suitable modifications as discussed in the following. Most planar flow is laminar and is expressed in gallons/minute/foot width or cubic meters/second/meter width.

Geotextile drains with Darcian flow are characterized hydraulically by manufacturers in terms of transmissivity based on ASTM D4716 (2013) or variations thereof (Koerner 2005) carried out under full-flow conditions. Transmissivity (m^3/sec-m or m^3/min-m) is measured at unit hydraulic gradient for a particular geotextile vs. normal stress (kPa), which takes into account the geotextile's compressibility under load and attendant variation in thickness. The standard is based on laminar (actually Darcian) flow tests and, although recognizing the possibility of transitional and turbulent flow in such conduits, stipulates that the test data for hydraulic transmissivity be obtained from the laminar test range. However, the laminar range is not generally reported; it is recommended here that such be required so that the laminar range of the test data can be known. Darcy's formula is then used to calculate the transmissivity required for the actual hydraulic gradient on the assumption that the hydraulic gradient equals the geotextile slope. Factors of safety and reduction factors (Koerner 2005) relate the required transmissivity to the ultimate transmissivity.

It is also important to note that the D4716 tests are for pressurized, full-conduit, constant flow. The applicability of these methods to spatially varied, free-surface flow applications must be considered. A useful criterion is derived by Graber (2007b) and leads to the following when the inflow is uniform (constant inflow per unit length); the first of the following inequalities applies subject to the second and third of the following inequalities:

$$\frac{d\frac{y}{b}}{d\frac{x}{L}} \geq 0, \quad \frac{S_o L}{b} \geq \frac{S_f L}{b}, \quad S_o L/b \gg 2q_L^2/(gpb^3) \qquad (6\text{-}16)$$

in which x = distance along the drain in the direction of flow; y = flow depth; L = total conduit length; S_o = conduit slope; S_f = friction slope; q_L = flow per unit width at the downstream end of the drain; g = gravitational acceleration; and p = medium porosity.

This relationship indicates that if the drain is designed for a flow equal to or less than the flowing-full capacity of the drain (i.e., $S_o \geq S_f$), the tailwater does not submerge the crown of the drain, and $S_o L/b \gg 2q_L^2/(gpb^3)$, the water surface will remain within the drain. Thus, for Darcian flow in geotextiles, with qualifications just discussed, design can rationally be based on transmissivity measurements for full-conduit Darcian flow conditions and, where cross-sectional distortion is significant, matching normal stress. Darcy's formula is then used to extrapolate to other hydraulic gradients with the assumption that the full-conduit geotextile slope is equal to the extrapolated hydraulic gradient. This method is valid for open-channel, spatially increasing flow design provided that the Eq. 6-16 inequalities are satisfied. For this method, some conservatism results from the lesser wetted perimeter compared with the full-conduit test measurements. Tailwater must be at or below the downstream crown, and conservatism is gained to the extent that tailwater is below the downstream crown. Conservatism is also gained if extrapolating to smaller normal-stress conditions for which there is less cross-sectional distortion.

For geotextile drains with Darcian, homogeneous in-plane flow under more general conditions when Eq. 6-16 inequalities are not satisfied (or even if they are), but subject to the same restrictions as discussed previously for aggregate drains, a rational design can be based on the same method as given for aggregate drains (Graber 2007b). The tailwater must be below that corresponding to the Numerov boundary condition. Equations 6-11 and Fig. 6-6 then apply. These relationships are not limited by the slope and other restrictions mentioned in the previous paragraph and can enable greater design economy. The relationship between the measured full-flow transmissivity and other variables, including R', are the same as discussed for aggregate drains. The factor of safety should selected based on the accuracy to which the hydrology (flow rate) is known, the practical importance of inaccuracies in assumed hydrology, and the accuracy to which drain-medium conductivity is known. Other considerations also pertain as discussed by Koerner (2005) in the context of flow-rate reduction factors.

Solutions for geotextile drains with tailwater higher than that corresponding to the Numerov condition can be obtained for known flow distribution and downstream flow depth using the numerical, spatially varied flow computational procedures such as those described under "Numerical Formulation" of Graber (2004). In doing so, the friction slope S_f should be taken as i, and the velocities should be taken as the mean interstitial velocity given by $Q/(Ap)$, with Q = flow rate, A = wetted cross-sectional area, and p = medium porosity. Should the flow be in transitional or turbulent regimes, considerations similar to those described previously in connection with aggregate drains apply.

Additional perspective regarding these methods and their relation to other references is given in Graber (2007b).

6.8.6 Hydraulics of Geonets and Geocomposite Drains. The following should be considered when geonets or geocomposite drains are utilized:

- Geonets and geocomposite drains generally have more complex flow patterns than pipes, thus contributing to higher flow energy losses.
- Certain designs are more hydraulically efficient, thus providing higher flow rates per unit width or thickness.
- The outer geotextile wrapping on geocomposite drains may affect hydraulic properties.
- Most geonets and geocomposites drains react noticeably to applied earth loads.

Flow in geonets and geocomposite drains is commonly turbulent. Graber (2007b) discusses means for determining the existence of turbulent flow in geocomposite drains, which should be used as necessary. Koerner (2005) suggests that design be based on full-conduit, constant-flow test measurements of the flow rate per unit width (m^3/min-m) for a particular material versus normal stress (kPa) and at different hydraulic gradients. The ultimate flow is then taken from the data at the design normal stress and geosynthetic slope equal to the test hydraulic gradient (interpolated as necessary), and then reduction factors and factors of safety applied. For both the laminar- and turbulent-flow cases, the assumption that the hydraulic gradient equals the geosynthetic slope is equivalent to the uniform-flow assumption in a just-full conduit based on the downstream-most flow.

The adequacy of these assumptions for spatially varied, free-surface flow applications can be tested using the inequalities in Eq. 6-16 and the related discussion, which apply equally to laminar, transitional, or turbulent flow (the flow regime directly affects S_f). With $p = 1$ Eq. 6-16 applies to polymer-core geocomposite drains of rectangular cross section, while geonets will generally have $p < 1$. Graber (2007b) gives an example of the application of Eq. 6-16 to a landfill geonet. For geonets, Eqs. 6-14 and 6-15 with their associated discussion may also be useful.

For the methods discussed in the previous two paragraphs, some conservatism results from the lesser wetted perimeter compared with the full-conduit test measurements. Tailwater must be at or below the downstream crown, and conservatism is gained to the extent that tailwater is below the downstream crown. Conservatism is also gained if extrapolating to smaller normal-stress conditions for which there is less cross-sectional distortion.

An additional consideration concerns the possibility that the distortion of the longitudinal velocity profile due to spatial inflow can result in an increase in effective wall friction (Graber 2004). That can make the roughness coefficient determined under spatially increasing flow conditions greater than that resulting from constant-flow tests. Aided by the seminal work of McNown (1954), Graber (2007b) defines the following criterion for minimal increases in effective wall friction compared to constant flow:

$$N_o > 11 \tag{6-17}$$

in which N_o = total number of columns of holes. For fabric-covered geocomposite drains without slots or holes, N_o can be approximated by

$$N_o = sf_o \frac{yL}{\frac{\pi}{4}(AOS)^2} \tag{6-18}$$

in which s = number of sides effectively receiving drainage, f_o = fraction of side area open for drainage, AOS = apparent opening size, and y = an appropriate flow depth.

For the design method for turbulent flow in geonets and geocomposite drains described previously with qualifications, to the extent that the tailwater elevation is below the downstream crown a factor of safety is introduced. For cases in which Eq. 6-16 cannot be satisfied, the design should be accomplished by numerical, spatially varied flow methods, such as discussed under "Numerical Formulation" in Graber (2004). It should be recognized that for such flows, the maximum depth can occur at an intermediate location along the channel (Graber 2004).

For highway edge drains, it can be important to design for a maximum free-surface depth and freeboard within the edge drain (Dempsey 1988, NCHRP 2004). Rather than the method discussed at the beginning of this subsection, that may be accomplished more economically by an alternative method. The alternative method discussed here is Graber's (2007b) modification/correction of methods presented by FHWA (1990, 2002) and NCHRP (2004). Graber (2007b) uses Dempsey's (1988) experimental data to properly estimate a constant-flow friction factor in the form of Manning's roughness coefficient; Table 6-2 presents the n values thus calculated. Manning's equation can be used to calculate the friction slope in the numerical formulation, using properly determined values of Manning's n such as developed for Table 6-2. Graber (2007b) presents an NCHRP (2004) example for a high-column edge drain using the numerical computational procedures for spatially varied, open-channel flow described under "Numerical Formulation" of Graber (2004) and also incorporating use of Eq. 6-18. That example is compared to the corresponding NCHRP (2004) result.

Additional perspective regarding these methods and their relation to other references is given in Graber (2007b).

Table 6-3 presents Manning's n values extrapolated from Table 6-2 for current products. Where n values on Table 6-3 are the same as those on Table 6-2, it is because the core specifications and hence surface roughness characteristics (roughness height and pattern) are the same and the geocomposite thicknesses are equal or the thicknesses in Table 6-3 are slightly greater (Marlow 2011a). In the latter case, the Table 6-3 values should be conservative. Caution should be exercised in extrapolating to other surface roughness characteristics or thicknesses.

6.8.7 Hydraulics of Soil Layer Consolidation Drains. Flow within consolidation drains is proportional to the cross-sectional area and the hydraulic conductivity of the drain fill material in the case of sand drains, and flow energy is dissipated by moving the water column upward. The vertical drain must not develop significant restrictions to flow within the drain. Head losses in vertical drains are determined by

$$h_d = \frac{q_d d}{2kA} \tag{6-19}$$

Table 6-2. Description of Geocomposite Fin-Drain Materials Modified from Dempsey (1988) and FHWA (2002) (Graber 2007b)

Core Data				Fabric Data			
Fin-Drain Material	Structural	Material	Thickness mm (in.)	Material	Fabrication	Core Attachment	Calculated Manning's n value
Hitek 20[a]	Cuspated	High-density polyethylene	20 (0.78)	Polypropylene	Nonwoven	Loose Wrapped	0.107
Hitek 40[a]	Cuspated	High-density polyethylene	40 (1.57)	Polypropylene	Nonwoven	Loose Wrapped	0.0935
Akwadrain[a]	Cuspated	High-density polyethylene	25 (1.00)	Polypropylene	Nonwoven	Adhesive Bond One Side, Loose One Side	0.188
Miradrain 6000	Dimpled Sheet	High-impact polysterene	10 (0.39)	Polypropylene	Nonwoven, Needle Punched	Adhesive Bond Two Sides	0.0771
Hydraway	Columns	Linear low-density polyethylene	25 (1.00)	Polypropylene	Nonwoven, Needle Punched, Calendered	Adhesive Bond to Columns, Heat Bond Backing	0.0691
J-Drain	Net	Low-density polyethylene	6.4 (0.25)	Polypropylene	Nonwoven	Linear Adhesive Bond Line Both Sides	0.0772

[a]Two-sided flow, with centrally located impermeable core membrane

Table 6-3. Description of Geocomposite Drainage Products Extrapolated from Table 6-2 (Marlow 2011a)

Core Data				Fabric Data			Estimated Manning's *n* value
Fin-Drain Material	Structural	Material	Thickness mm (in.)	Material	Fabrication	Core Attachment	
AWD Site Drain Strip 6000[a]	Dimpled Sheet	High-impact polysterene	25.4 (1.0)	Polypropylene	Nonwoven, Needle Punched	Adhesive Bond Dimple Side, Loose Flat Side	0.188
Contech Strip Drain[a]	Dimpled Sheet	High-impact polysterene	25.4 (1.0)	Polypropylene	Nonwoven, Needle Punched	Adhesive Bond Dimple Side, Loose Flat Side	0.188
Hydraway 2000[a]	Columns	Low-density polyethylene	25.4 (1.0)	Polypropylene	Nonwoven, Needle Punched	Adhesive Bond to Columns, Heat Bond Backing	0.0691
AWD Site Drain Sheet 180	Dimpled Sheet	High-impact polysterene	11 (0.44)	Polypropylene	Nonwoven, Needle Punched	Adhesive Bond Dimple Side	0.0771
Contech C-Drain 15K	Dimpled Sheet	High-impact polysterene	11 (0.44)	Polypropylene	Nonwoven, Needle Punched	Adhesive Bond Dimple Side	0.0771
CCW Miradrain 6000	Dimpled Sheet	High-impact polysterene	10 (0.39)	Polypropylene	Nonwoven, Needle Punched	Adhesive Bond Dimple Side	0.0771
JDR J-Drain 300	Net	High-density polyethylene	6.4 (0.25)	Polypropylene	Nonwoven, Needle Punched	Heat-Fused One Side	0.0772

[a]Two-sided flow

where

h_d = total head loss in drain, feet (meters);
q_d = groundwater discharge for each drain, cubic feet per second (cubic meters per second);
d = depth of the drain, feet (meters);
k = permeability coefficient of drain backfill, feet per second (meters per second) ; and
A = cross-sectional area of the drain per running foot, square feet per foot (square meters per meter).

Note that higher drain backfill permeability and larger cross-sectional area reduce head loss proportionally. As consolidation progresses, the density of the drain backfill material may increase, and an accompanying decrease in permeability may occur.

Geocomposite vertical drains may be evaluated in a fashion similar to sand drains. In determining head loss, the hydraulic transmissivity (flow rate potential) of the wick drain (determined by laboratory test) is substituted for the permeability coefficient and the cross-sectional area in Eq. 6-19.

CHAPTER 7
STRUCTURAL CONSIDERATIONS

7.1 LOADING

7.1.1 General. The loads that may be applied to an urban subsurface drainage system are categorized as either dead loads or live loads. Equipment live loads may be critical during construction operations.

Pipe and geocomposite systems are designed to carry live loads and dead loads. Geomembrane, geotextile, and aggregate systems, by their nature, normally do not have to be designed for structural strength to carry live loads or dead loads but must be designed for stability and possible shear forces during construction.

7.1.2 Dead Loads and Live Loads. Dead loads may include earth loads, weight of pipe, weight of fluid in the system, building foundation loads, and surcharge loads. The primary dead load, which usually governs, is the earth load. The magnitude of the earth load depends on the unit weight of the soil and depth of the fill.

Live loads are imposed by highway vehicles, trains, aircraft, and construction equipment and are typically taken as the AASHTO H-20, HS-20, or alternate interstate loading for highways, and the Cooper E 80 loading for railroads. The magnitudes of these loads are determined by design methods found in the following references: American Association of State of Highway and Transportation Officials (AASHTO HB-17) for highway loadings; American Concrete Pipe Association (2004); American Railway Engineering and Maintenance-of-Way Association (AREMA 2010), for railroad loadings; National Clay Pipe Institute (2006); National Corrugated Steel Pipe Association (NCSPA 2008); Spangler and Handy (1982); USDA, SCS (1958, 1980); and Federal Aviation Administration (2006), for aircraft loadings.

7.1.3 Construction Loads. Construction loads from heavy construction equipment traveling over or across an installed drain system may create load concentrations in excess of design loads that may displace and damage the system. Such locations should be evaluated by the engineer to determine if displacement and damage may occur. If necessary, crossing location requirements should be detailed in the construction contract documents. Although crossings for pipe systems should be evaluated on a case-by-case basis, a common crossing consists of a temporary earth fill constructed to an elevation of at least 3 feet (1 m) over the top of the system and to a width sufficient to prevent lateral displacement of the system.

7.2 EMBEDMENT

The structural performance of buried pipe depends on the interaction between the soil and the pipe. Therefore, the pipe embedment must be selected for structural and drainage characteristics. Structural characteristics of the embedment include consideration of the dimensions of the embedment around the pipe; the type of soil, density, potential for live loads, and compaction of the embedment, native soil, and fill; the depth of burial of the pipe; and the height and characteristics of the water table. The required dimensions and type of soil, density, and compaction of the embedment depend on the pipe stiffness.

Flexible pipes, such as plastic and corrugated metal, use embedment materials to transfer vertical loads into the adjacent soil. Rigid pipes, such as concrete and clay, transfer vertical loads directly into the bedding with minimal load transfer into the adjacent soil. Therefore, the required structural characteristics of the embedment vary with the type of pipe and should be determined in accordance with 7.3.

The type of equipment to be used to compact the embedment and fill should be evaluated by the engineer to determine if pipe damage and displacement may result from its use. If necessary, equipment limitations should be detailed in the construction contract documents.

7.3 PIPE DESIGN

Several pipe design methods exist. The appropriate method depends on the pipe application or end use, the type of pipe material, and possibly the project authority or owner. Acceptable design methods for pipe projects within the scope of this practice are presented for each pipe material in 7.3.1 through 7.3.4.

The trench bottom should have a 120° semicircular groove cut into it to cradle the pipe and provide structural support. The groove should be cut, not pressed, in the trench bottom. The groove can also be a 90° V-groove. The groove can be cut into the bottom of the trench by the pipe feeding boot (i.e., install the pipe by feeding it through the boot or shoe of the trencher) or by the cutting blades on the digging apparatus. If the bottom of the trench is filled with fine gravel that surrounds the pipe, the groove is not necessary.

7.3.1 Corrugated Metal Pipe. Service load and load factor design methods and embedment requirements are presented in Sections 3 and 12 of *Standard Specifications for Highway Bridges* (AASHTO HB-17); in *Corrugated Steel Pipe Design Manual* (NCSPA 2008); and in ASTM A796/A796M, *Standard Practice for Structural Design of Corrugated Steel Pipe, Pipe-Arches, and Arches for Storm and Sanitary Sewers and Other Buried Appurtenances.*

More recent and other important references include ASCE Manual of Practice No. 119, *Buried Flexible Steel Pipe: Design and Structural Analysis* (2009); Smith and Watkins (2004), "The Iowa Formula: Its Use and Misuse when Designing Flexible Pipe"; and Watkins, et al. (2010), "Pipe Zone Bedding and Backfill: A Flexible Pipe Approach."

7.3.2 Thermoplastic Pipe. Sources of design information are listed in the following. Manufacturers' literature should also be checked for recommended design methods and embedment requirements.

- *Handbook of PVC Pipe: Design and Construction* (Uni-Bell PVC Pipe Association 2001) is a unified source for PVC pipe design and construction.
- ASCE Manual of Practice No. 60, *Gravity Sanitary Sewer Design and Construction* (ASCE 2007) presents design methods and requirements for all thermoplastic pipes.
- NCHRP Report 225, "Plastic Pipe for Subsurface Drainage of Transportation Facilities" (Chambers et al. 1980), presents a review of thermoplastic pipe design methods.
- *The Complete Corrugated Polyethylene Pipe Design Manual and Installation Guide*, published by the Corrugated Polyethylene Pipe Association (2005), presents information on design of corrugated polyethylene pipe.
- Sections 3 and 17 of *Standard Specifications for Highway Bridges* (AASHTO HB-17) presents service load and load factor design methods and embedment requirements for some PVC and PE pipe products.
- ASTM D2321 *Standard Practice for Underground Installation of Thermoplastic Pipe for Sewers and Other Gravity-Flow Applications*.
- Rahman, S. and Watkins, R. K. (2004). "Longitudinal Mechanics of Buried Thermoplastic Pipe: Analysis of PVC Pipes of Various Joint Types." *Proceedings of the Pipeline Division Specialty Conference*, Aug. 21-24, 2005, Houston, Texas, 1101-1116.
- More recent and other important references for corrugated plastic pipe include Smith and Watkins (2004), "The Iowa Formula: Its Use and Misuse when Designing Flexible Pipe"; Watkins (2004), "Pipe and Soil Mechanics for Buried Corrugated HDPE Pipe"; Watkins, et al. (2010), "Pipe Zone Bedding and Backfill: A Flexible Pipe Approach."

7.3.3 Precast Concrete Pipe. Service load and load factor design methods and embedment requirements are presented in Sections 3 and 9 of *Standard Specifications for Highway Bridges* (AASHTO HB-17). The American Concrete Pipe Association's *Concrete Pipe Design Manual* (2004) presents data and information on the design of concrete pipe systems. The design manual is a companion volume to the *Concrete Pipe Handbook* (American Concrete Pipe Association 1998), which provides an up-to-date compilation of the concepts and theories that form the basis for the design and installation of precast concrete pipe. ASCE 15-98, *Standard Practice for Direct Design of Buried Precast Concrete Pipe Using Standard Installations* also contains valuable information on this subject. ASCE 26-97, *Standard Practice for Direct Design of Buried Precast Concrete Box Sections*; ASCE 27-00, *Standard Practice for Direct Design of Precast Concrete Pipe for Jacking in Trenchless Construction*; and ASCE 28-00, *Standard Practice for Direct Design of Precast Concrete Box Sections for Jacking in Trenchless Construction* are similar references on precast concrete pipe design.

7.3.4 Clay Pipe. Service load design methods and embedment requirements are presented in *Clay Pipe Engineering Manual* (National Clay Pipe Institute 2006).

7.4 OTHER SYSTEMS

Geomembranes, geotextiles, geonets, and aggregate systems, by their nature, do not have to be designed for structural strength to carry live loads or dead loads. These systems, however, must be designed for stability, for possible shear forces from construction operations, and for the ability to maintain flow capacities after placement. Geomembranes, geotextiles, and geonets should have sufficient strength to prevent tearing during construction and operation. Maintenance of flow capacities of geotextiles and geonets in relation to loading is discussed in 6.8.5 and 6.8.6, respectively. Koerner discusses "survivability" requirements for geomembranes (Koerner 2005, Section 5.2) and geotextiles (Koerner 2005, Sections 2.1.2 and 2.11.2).

Geocomposites should have sufficient crush strength to withstand construction loading and compaction stresses. For roadways, Christopher and Zhao (2001) state that these typically range from 35 psi (240 kPa) to 70 psi (480 kPa) beneath the base coarse layer and 200 psi (1380 kPa) beneath the asphalt. FHWA (Holtz et al. 1998) states that ultimate strength should be based on minimum average roll values (MARV) as defined in ASTM D4759 *Standard Practice for Determining the Specification Conformance of Geosynthetics*, and that ultimate strength values for quality assurance (QA) purposes may be determined according to ASTM D4595 *Standard Test Method for Tensile Properties of Geotextiles by the Wide-Width Strip Method*.

Geocomposites should also be of such shape and have sufficient strength to prevent major deformations of the geotextile into the core that will reduce flow capacity under load. Maintenance of flow capacity of geocomposites in relation to loading is discussed in 6.8.6. For pavement applications, geocomposite drains should have sufficient strength to withstand a long-term compressive force of 20 psi (138 kPa) normal to the plane of the drain or twice the active soil pressure at the drain depth, whichever is greater. For other applications, geocomposite drains must be evaluated against the anticipated sustained dead loads to ensure that the applied loads do not exceed the proven allowable long-term load level to prevent creep deformations of the polymer materials, which would cause reduction in flow capacity.

Where repetitive (continuous or dynamic) loads are expected, a factor of safety should be applied to short-term load tests to prevent major deformations or collapse associated with time-dependent compressive behavior. For example, Christopher and Zhao (2001) note that dynamic traffic loading can induce significant creep deformation and potential geocomposite collapse. FHWA (Holtz et al. 1998) recommends that creep testing be conducted in accordance with ASTM D5262, *Standard Test Method for Evaluating the Unconfined Tension Creep and Creep Rupture Behavior of Geosynthetics* and that for confined creep testing (for which no standardized test exists) project-specific or representative backfill material and typical placement and confinement conditions should be used in the testing.

American Wick Drain (AWD) has provided material for informational purposes based on its internal test results, analysis, and experience; although not meant to be taken as statements of absolute fact based on extensive testing, the information is very useful and provided here. AWD has conducted creep testing on several of its products using ASTM D5262, modified for confined compressive creep testing rather than tension creep (no standardized test method currently exists for compressive creep testing) (Marlow 2011b). The company has tested three different plastic types in various dimpled core configurations to ascertain how they are influenced by compressive creep. The plastic types tested are high-impact polystyrene (HIPS), polypropylene (PP), and high-density polyethylene (HDPE). The results indicate that for cores designed to meet the same compressive strength (also referred to as initial crush strength, "compressive strength" as used in ASTM D1621, which defines compressive strength as the stress at the yield point), HIPS core products exhibit the best creep resistance, PP products exhibit good creep resistance, and

HDPE products exhibit poor creep resistance. Based on AWD's results, factor of safety (FS) = 3 would likely be adequate for the company's HIPS core products, FS = 4 would likely be adequate for PP core products, and FS = 5 or greater would likely be required for HDPE products (not currently manufactured by AWD in part for this reason). Thus, a single FS does not apply to all geocomposite drainage products based on initial crush strength as the core plastic and product design (dimples, cones, configurations, etc.) can play a large role in compressive creep results. The AWD compressive creep testing has been based on long-term static loading conditions.

For additional perspective, FHWA (Holtz et al. 1998) recommends controlled laboratory creep tests for a minimum duration of 10,000 hours, then extrapolated to the design life (typically 75 years or more). FHWA (Holtz et al. 1998) gives factors of safety (which they refer to as creep reduction factors) for different polymer types of 2.5 to 2.0 for polyester, 5.0 to 4.0 for polypropylene, and 5.0 to 2.5 for polyethylene, applied to the average ultimate tensile strength. ASTM D4716 (section X2) addresses long-term static loading in the context of hydraulic testing.

ASCE (2006) at 8.3.6 gives examples of long-term creep test results for geocomposite surface drains based on a modification of ASTM D1621, *Standard Test Method for Compressive Properties of Rigid Cellular Plastics*, and provides additional perspective.

CHAPTER 8
MATERIALS

8.1 PIPE

When specified, pipe should conform to the material and manufacturing requirements of the American Society for Testing and Materials (ASTM), American Association of State Highway and Transportation Officials (AASHTO), or American National Standards Institute/American Water Works Association (ANSI/AWWA) standards referenced. The following lists of standards are for products commonly accepted and used on current projects; it is not intended that these listings restrict the use of new products that may be developed and found satisfactory.

8.1.1 Concrete Pipe. Reinforced and nonreinforced concrete pipe are used for gravity flow systems. Reinforced concrete pressure pipe and prestressed concrete pressure pipe are used for pressure flow systems. Concrete fittings and appurtenances, such as wyes, tees, and manhole sections, are generally available. Several jointing methods are available depending on the tightness required and the operating pressure.

Several mechanical processes are used in the manufacture of concrete pipe. These processes use various techniques, including centrifugation, vibration, packing, and tamping to consolidate the concrete in forms. Gravity and pressure concrete pipe may be manufactured to any reasonable strength requirement by varying the wall thickness, concrete strength, and quantity and configuration of reinforcing steel or prestressing elements.

Gravity Flow Applications. Concrete pipe is specified by nominal diameter, type of joint, and D-load strength or reinforcement requirements. The product should be manufactured in accordance with one or more of the following standard specifications:

- ASTM C14/C14M, *Nonreinforced Concrete Sewer, Storm Drain, and Culvert Pipe.*
- ASTM C76/C76M/AASHTO M170/M170M, *Reinforced Concrete Culvert, Storm Drain, and Sewer Pipe,* covers reinforced concrete pipe intended for the conveyance of sewage, industrial wastes, and stormwater.
- ASTM C118/C118M, *Concrete Pipe for Irrigation or Drainage,* covers concrete pipe intended for the conveyance of water with specified working pressures including hydraulic transients.
- ASTM C361/C361M, *Reinforced Concrete Low-Head Pressure Pipe,* covers reinforced concrete pipe conduits with low internal hydrostatic heads generally not exceeding 125 ft (38 m).
- ASTM C412/C412M/AASHTO M178/M178M, *Concrete Drain Tile*, covers concrete drain tile with internal diameters from 4 to 36 in. (100 to 900 mm).
- ASTM C444/C444M/AASHTO M175/M175M, *Perforated Concrete Pipe*, covers perforated concrete pipe intended for underdrainage.
- ASTM C478/C478M, *Precast Reinforced Concrete Manhole Sections*, covers products used for the assembly and construction of circular vertical precast reinforced concrete manholes and structures used in sewer, drainage, and water works.
- ASTM C505/C505M, *Nonreinforced Concrete Irrigation Pipe with Rubber Gasket Joints*, covers pipe for the conveyance of water with specified working pressures.
- ASTM C506/C506M/AASHTO M206/M206M, *Reinforced Concrete Arch Culvert, Storm Drain, and Sewer Pipe*, covers reinforced concrete arch-shaped concrete pipe for the conveyance of sewage, industrial wastes, and storm water and for the construction of culverts.
- ASTM C507/C507M/AASHTO M207/M207M), *Reinforced Concrete Elliptical Culvert, Storm Drain, and Sewer Pipe*, covers reinforced elliptical concrete pipe for the conveyance of sewage, industrial wastes, and storm water and for the construction of culverts.
- ASTM C654/C654M/AASHTO M176/M176M, *Porous Concrete Pipe*, covers porous nonreinforced concrete pipe for use in underdrains.
- ASTM C655/C655M/AASHTO M242/M242M, *Reinforced Concrete D-Load Culvert, Storm Drain and Sewer Pipe*, covers reinforced concrete pipe for specific D-loads and intended for the conveyance of sewage, industrial wastes, and stormwater.
- AASHTO M259/M259M, *Precast Reinforced Concrete Box Sections for Culverts, Storm Drains, and Sewers*, covers single-cell precast reinforced concrete box sections with less than 2 ft (0.6 m) of earth cover when subjected to highway loads.
- AASHTO M273/M273M, *Precast Reinforced Concrete Box Sections for Culverts, Storm Drains, and Sewers with Less Than 2 Feet (0.6 m) of Cover Subjected to Highway Loading*, covers box sections with less than 2 ft (0.6 m) of earth cover subjected to highway loading intended for the construction of culverts and the conveyance of stormwater, industrial wastes, and sewage.
- ASTM C985/C985M, *Nonreinforced Concrete Specified Strength Culvert, Storm Drain, and Sewer Pipe*, covers nonreinforced concrete pipe designed for specified strengths.
- ASTM C1433/C1433M, *Precast Reinforced Concrete Monolithic Box Sections for Culverts, Storm Drains, and Sewers*, covers single-cell precast reinforced concrete box sections cast monolithically and proposed for use in the construction of culverts and for conveyance of stormwater, industrial wastes, and sewage.

Pressure Flow Applications. Concrete pressure pipe is specified to provide custom designs based on specific conditions of service. The product should be designed and manufactured in accordance with one or more of the following standards:

- ANSI/AWWA C300, *Reinforced Concrete Pressure Pipe, Steel Cylinder Type*, covers reinforced concrete cylinder pipe from 30- to 144-in. (760- to 3,660-mm) diameters and larger.
- ANSI/AWWA C301, *Prestressed Concrete Pressure Pipe, Steel Cylinder Type for Water and Other Liquids*, covers prestressed concrete cylinder pipe from 16- to 144-in. (410- to 3,660-mm) diameters and larger. Pipe larger than 250 in. (6,250 mm) in diameter has been produced.
- ANSI/AWWA C302, *Reinforced Concrete Pressure Pipe, Noncylinder Type*, covers reinforced concrete noncylinder pipe from 12- to 144-in. (300- to 3,660-mm) diameters and larger. Because it does not contain a watertight membrane (steel cylinder), reinforced concrete noncylinder pipe is limited to internal pressures of 55 psi (385 kPa) or less.
- ANSI/AWWA C303, *Concrete Pressure Pipe, Bar-Wrapped, Steel-Cylinder Type*, covers concrete pressure pipe with a steel cylinder that is wrapped with mild steel bar reinforcement from 10- to 72-in. (250- to 1,830-mm) diameters. Larger pipe sizes and pipe accommodating higher pressures have been manufactured based on the concepts of this standard.

8.1.2 Thermoplastic Pipe. Thermoplastic pipe materials include various plastics that can be repeatedly softened by heating and hardened by cooling through a temperature range characteristic for each specific plastic and in the softened state can be shaped by molding or extrusion. Generally, thermoplastic pipe materials are limited to acrylonitrile-butadiene-styrene (ABS), polyethylene (PE), and polyvinyl chloride (PVC). Thermoplastic pipes are produced in various shapes and dimensions. Pipe properties can be modified by changing the wall thickness or profile for both pressure and nonpressure applications.

8.1.2.1 Acrylonitrile-Butadiene-Styrene Pipe. Acrylonitrile-butadiene-styrene (ABS) pipe is manufactured by extrusion of ABS material and is limited to gravity flow applications. ABS composite pipe is manufactured by extrusion of ABS material with a series of truss annuli that are filled with filler material such as lightweight Portland cement concrete. ABS fittings are available for the product. The jointing systems available include elastomeric gasket joints and solvent cement joints.

Gravity Flow Applications. ABS pipe should be manufactured in accordance with one of the following standard specifications:

- ASTM D2680/AASHTO M264, *Acrylonitrile-Butadiene-Styrene (ABS) and Poly (Vinyl Chloride) (PVC) Composite Sewer Piping*, covers ABS or PVC composite pipe, fittings, and a joining system for nonpressure systems.
- ASTM D2751, *Acrylonitrile-Butadiene-Styrene (ABS) Sewer Pipe and Fittings*, covers ABS pipe and fittings from 3- to 12-in. (75- to 300-mm) diameter.

8.1.2.2 Polyethylene Pipe. Polyethylene (PE) pipe is used for both gravity and pressure flow systems. PE pipe is manufactured by extrusion of PE plastic material. PE refers to the base compound, whereas finished polymers may be referred to as LLDPE (linear low-density polyethylene) or HDPE (high-density polyethylene). PE pipe is specified by material designation, nominal diameter (inside or outside), standard dimension ratios, ring stiffness, and type of joint. PE fittings are available as well as butt fusion welding.

Gravity Flow Applications. PE pipe for gravity flow applications should be manufactured in accordance with one or more of the following standard specifications:

- ASTM F405, *Corrugated Polyethylene (PE) Pipe and Fittings*, covers pipe from 3- to 6-in. (75- to 150-mm) diameters.
- AASHTO M252, *Corrugated Polyethylene Drainage Pipe*, covers pipe from 3- to 10-in. (75- to 250-mm) diameters.
- ASTM F667, *Large Diameter Corrugated Polyethylene Pipe and Fittings*, covers pipe from 8- to 24-in. (203- to 610-mm) diameters.
- AASHTO M294, *Corrugated Polyethylene Pipe 300- to 1500-mm (12- to 60-in.)*.
- ASTM F810, *Smoothwall Polyethylene (PE) Pipe for Use in Drainage and Waste Disposal Absorption Fields*, covers smoothwall PE pipe, including coextruded, perforated, and nonperforated from 3- to 6-in. (75 to 150-mm) diameters.
- ASTM F894, *Polyethylene (PE) Large Diameter Profile Wall Sewer and Drain Pipe*, covers profile wall PE pipe from 10- to 120-in. (250- to 3,050-mm) diameters for gravity flow applications.
- AASHTO MP 20-10, *Steel-Reinforced Polyethylene (PE) Ribbed Pipe, 300- to 900-mm (12- to 36-in.) Diameter*, covers a class of pipe containing steel reinforcing rods that are molded into the pipe ribs. This imparts higher strength and allows the pipe to be designed as nonflexible, especially for larger diameters.
- ASTM F2562/F2562M, *Steel Reinforced Thermoplastic Ribbed Pipe and Fittings for Non-Pressure Drainage and Sewerage*, covers steel reinforced thermoplastic pipe and fittings of nominal sizes 8 in. (200 mm) through 120 in. (3000 mm), intended for use in underground applications where soil provides support for their flexible walls, used for gravity flow and nonpressure applications (such as storm sewers, drainage pipes, etc.)

Pressure Flow Applications. PE pressure pipe should be manufactured in accordance with one or more of the following standard specifications:

- ASTM D2239, *Polyethylene (PE) Plastic Pipe (SIDR-PR) Based on Controlled Inside Diameter*, covers PE pipe made in standard thermoplastic pipe dimensions and pressure rated for water.
- ASTM D3035, *Polyethylene (PE) Plastic Pipe (DP-PR) Based on Controlled Outside Diameter*, covers PE pipe made in standards thermoplastic pipe dimensions based on outside diameter, and pressure rated for water.

8.1.2.3 Polyvinyl Chloride Pipe. PVC pipe is used for both gravity and pressure flow systems. PVC pipe is manufactured by extrusion of the material. Higher-strength PVC composite pipe is manufactured by extrusion of PVC material with a series of truss annuli that are filled with material such as lightweight Portland cement concrete. PVC pipe is specified by nominal diameter, dimension ratio, pipe stiffness, and type of joint. PVC pressure and nonpressure fittings are available. Joints are typically solvent welded or gasketed depending on the application and diameter.

Gravity Flow Applications. PVC pipe for gravity flow applications should be manufactured in accordance with one or more of the following standard specifications:

- ASTM D2680/AASHTO M264, *Acrylonitrile-Butadiene-Styrene (ABS) and Poly (Vinyl Chloride) (PVC) Composite*

Sewer Piping, covers ABS or PVC composite pipe, fittings, and a joining system for nonpressure sanitary sewer and storm drain systems.

- ASTM D2729, *Poly (Vinyl Chloride) (PVC) Sewer Pipe and Fittings*, covers material and test requirements for PVC pipe and fittings for sewer and drain pipe. Standard perforations available only in 4-in. (100-mm) diameter pipe.
- ASTM D3034, *Type PSM Poly (Vinyl Chloride) (PVC) Sewer Pipe and Fittings*, covers material and test requirements for PVC pipe and fittings for sewer pipe systems.
- ASTM F679, *Poly (Vinyl Chloride) (PVC) Large-Diameter Plastic Gravity Sewer Pipe and Fittings*, covers material and test requirements for PVC gravity sewer pipe and fittings from 18- to 48-in. (450- to 1,200-mm) diameters, with integral bell elastomeric seal joints and smooth inner walls.
- ASTM F758, *Smooth-Wall Poly (Vinyl Chloride) (PVC) Plastic Underdrain Systems for Highway, Airport, and Similar Drainage*, covers material and test requirements for smoothwall pipe and fittings for PVC underdrains from 4- to 8-in. (100- to 200-mm) diameters with perforated or nonperforated walls for use in subsurface drainage systems.
- ASTM F794, *Poly (Vinyl Chloride) (PVC) Profile Gravity Sewer Pipe and Fittings Based on Controlled Inside Diameter*, covers materials and test requirements for PVC gravity sewer profile pipe and fittings with integral bell and elastomeric seal joints.
- ASTM F949, *Poly (Vinyl Chloride) (PVC) Corrugated Sewer Pipe with a Smooth Interior and Fittings*, covers materials and test requirements for PVC pipe and fittings from 4- to 48-in. (100- to 1,200-mm) diameters with corrugated outer wall fused to a smooth inner wall for sanitary and storm sewers and perforated and nonperforated pipe for subdrainage.

Pressure Flow Applications. PVC pressure pipe should be manufactured in accordance with one of the following standard specifications:

- ASTM D1785, *Poly (Vinyl Chloride) (PVC) Plastic Pipe, Schedules 40, 80, 120*, covers materials and test requirements for PVC pipe pressure rated for use with the distribution of pressurized liquids.
- ASTM D2241, *Poly (Vinyl Chloride) (PVC) Pressure Rated-Pipe (SDR Series)*, covers materials and test requirements for PVC pipe pressure rated for water.
- ANSI/AWWA C900, *Polyvinyl Chloride (PVC) Pressure Pipe and Fabricated Fittings, 4 in. Through 12 in. (100 mm Through 300 mm), for Water Transmission and Distribution*, covers materials and test requirements for PVC pipe pressure rated for water.
- ANSI/AWWA C905, *Polyvinyl Chloride (PVC) Pressure Pipe and Fabricated Fittings, 14 in. Through 48 in. (350 mm Through 1200 mm), for Water Transmission and Distribution*, covers materials and test requirements for PVC pipe pressure rated for water transmission in sizes ranging from 14- to 48-in. (350- to 1,200-mm) outside diameters and dimension ratios (DRs) of 14, 18, 21, 25, 26, 32.5, 41, and 51.

8.1.3 Metal Pipe. Corrugated metal pipe is fabricated from corrugated steel, aluminized steel, or aluminum sheets or coils. Corrugated metal pipe is specified by size, shape, wall profile, gauge or wall thickness, and coating or lining. Appurtenances, including tees, wyes, elbows, and manholes, are available. Corrugated metal pipe is limited to gravity flow applications.

Gravity Flow Applications. Corrugated metal pipe should be manufactured in accordance with one or more of the following standard specifications:

- ASTM A760/A760M/AASHTO M36, *Corrugated Steel Pipe, Metallic-Coated for Sewers and Drains*, covers metallic-coated corrugated steel pipe from 4- to 144-in. (100- to 3,600-mm) diameters.
- ASTM A761/A761M/AASHTO M167/M167M, *Corrugated Steel Structural Plate, Zinc-Coated, for Field-Bolted Pipe, Pipe Arches, and Arches*, covers corrugated steel structural plate, zinc-coated, used in the construction of pipe and other shapes for field assembly.
- ASTM A762/A762M/AASHTO M245, *Corrugated Steel Pipe, Polymer Precoated for Sewers and Drains.*
- ASTM A849, *Post-Applied Coatings, Pavings, and Linings for Corrugated Steel Sewer and Drainage Pipe*, covers post-applied coatings, pavings, and linings for corrugated steel pipe and corrugated steel structural plate pipe; pipe arches; and arches coated, paved, or lined with specified materials over either metallic coatings or metallic coatings with polymer coatings.
- AASHTO M190, *Bituminous Coated Corrugated Metal Culvert Pipe and Pipe*, covers bituminous-coated corrugated metal pipe and pipe arches intended to be used for the construction of metal culverts of the following types: fully bituminous-coated, half bituminous-coated with paved invert, fully bituminous-coated and paved-invert, fully bituminous-coated and 100% paved or lined.
- ASTM B745/B745M/AASHTO M196, *Corrugated Aluminum Pipe for Sewers and Drains.*
- AASHTO M274, *Steel Sheet, Aluminum Coated (Type 2), for Corrugated Steel Pipe*, covers steel sheet used in the fabrication of corrugated steel pipe used for drainage pipe and underdrains. The steel sheet is coated with commercially pure aluminum (referred to as Type 2) on continuous lines by the hot-dip process.

8.1.4 Vitrified Clay Pipe. Vitrified clay pipe (VCP) is manufactured from clays and shales and vitrified at high temperatures. VCP is available in standard and extra-strength classifications and is specified by nominal pipe diameter, strength, and type of joint. The product is limited to gravity flow applications.

Gravity Flow Applications. The product should be manufactured in accordance with one or more of the following standard specifications:

- ASTM C4, *Clay Drain Tile and Perforated Clay Drain Tile.*
- ASTM C700, *Vitrified Clay Pipe, Extra Strength, Standard Strength, and Perforated*, covers perforated and nonperforated pipe from 3- to 48-in. (76- to 1,220-mm) diameters in extra strength and standard strength.

8.2 CORRUGATED WALL STORMWATER COLLECTION CHAMBERS

Designed primarily for low profile subsurface storage and infiltration of urban runoff, corrugated wall stormwater collection chambers (ASTM F2418) are typically placed under paved surfaces in heavily developed areas. These devices are typically designed to interlock in rows with a manifold of connection pipes to transfer water from one chamber to another. Installations typically require a structural rock base with structural fill between the chambers to transfer load to the subbase. Infiltration rates depend on the rates of the underlying native soils. Storage is based on the sum of the chamber volume and structural rock void. Where used, design is based on regulatory guidelines, and typically pretreatment of runoff entering the system of chambers is required to prevent clogging and reduced infiltration rates.

8.3 OTHER MATERIALS AND PRODUCTS

Some national standard specifications for geocomposites, geonets, geomembranes, geotextiles, aggregates, and wick drains (also called chimney or vertical strip drains) are listed below. Numerous additional standards for geomembranes and geotextiles exist, which the reader can locate at the standards organizations.

- ASTM C136, *Standard Test Method for Sieve Analysis of Fine and Coarse Aggregates*, covers the determination of the particle size distribution of fine and coarse aggregates by sieving.
- ASTM D459, *Standard Test Method for Tensile Properties of Geotextiles by the Wide-Width Strip Method*, covers a procedure for determining the conformance of geosynthetic properties to standard specifications.
- ASTM D4716, *Standard Test Method for Determining the (In-plane) Flow Rate per Unit Width and Hydraulic Transmissivity of a Geosynthetic Using a Constant Head*, covers a procedure for determining the flow rate per unit width within the manufactured plane of geosynthetics under varying normal compressive stresses and a constant head.
- ASTM D4751, *Standard Test Method for Determining Apparent Opening Size of a Geotextile*, gives a test method used to determine the apparent opening size (AOS) of a geotextile by sieving glass beads through a geotextile.
- ASTM D4759, *Standard Practice for Determining the Specification Conformance of Geosynthetics*, covers the measurement of tensile properties of geotextiles using a wide-width strip specimen tensile method. This test method is applicable to most geotextiles that include woven fabrics, nonwoven fabrics, layered fabrics, knit fabrics, and felts that are used for geotextile application.
- ASTM D5199, *Standard Test Method for Measuring the Nominal Thickness of Geosynthetics*, covers the measurement of the nominal thickness of geotextiles, smooth-surfaced geomembranes, geonets, and geocomposite drainage products.
- ASTM D5262, *Standard Test Method for Evaluating the Unconfined Tension Creep and Creep Rupture Behavior of Geosynthetics,* gives a test method for use in determining the unconfined tension creep and creep rupture behavior of geosynthetics at constant temperature when subjected to a sustained tensile loading.
- ASTM D6088, *Standard Practice for Installation of Geocomposite Pavement Drains*, covers recommendations and identifies pertinent areas of consideration for the installation of buried geocomposite drains used for highway edge drains, underdrains, or other pavement drainage applications meeting the requirement of ASTM D7001.
- ASTM D6244, *Standard Test Method for Vertical Compression of Geocomposite Pavement Panel Drains*, covers vertical strain and core area change of geocomposite pavement drains, such as those included in ASTM D7001, under vertical compression.
- ASTM D6364, *Standard Test Method for Determining Short-Term Compression Behavior of Geosynthetics*, covers procedures for evaluation of the deformations of a geosynthetic or combination of geosynthetics (i.e., certain geocomposites) under short-term compressive loading.
- ASTM D6707, *Standard Specification for Circular-Knit Geotextile for Use in Subsurface Drainage Applications*, covers circular-knit geotextiles for use on the outside of perforated pipes and Class B geocomposites per ASTM D7001 in drainage applications.
- ASTM D6917, *Standard Guide for Selection of Test Methods for Prefabricated Vertical Drains (PVD)*, provides recommendations for the selection of appropriate test methods for prefabricated vertical geocomposite drains (sometimes referred to as wick drains) used in geotechnical engineering applications to provide consistency in data reporting.
- ASTM D6918, *Standard Test Method for Testing Vertical Strip Drains in the Crimped Condition*, gives a performance test that measures the effect crimping has on the ability of vertical strip drains to transmit water parallel to the plane of the drain.
- ASTM D7001, *Standard Specification for Geocomposites for Pavement Edge Drains and Other High-Flow Applications*, covers geocomposite drainage panels used in highway, turf, and other high-volume subsurface drainage applications. These products consist of a geotextile wrapped around a polymer core, a minimum of 25 mm (1 in.) thick and available in 150-mm (6-in.), 300-mm (12-in.), 450-mm (18-in.), and 600-mm (24-in.) widths.
- ASTM D7005, *Standard Test Method for Determining the Bond Strength (Ply Adhesion) of Geocomposites*, defines a procedure for comparing the bond strength or ply adhesion of geocomposites.
- ASTM D7273, *Standard Guide for Acceptance Testing Requirements for Geonets and Geonet Drainage Geocomposites*, covers guidelines for the acceptance testing frequency requirements for geonet and geonet drainage geocomposite materials, describing types of tests, test methods, and recommended verifications.
- ASTM D7498, *Standard Test Method for Vertical Strip Drains Using a Large-Scale Consolidation Test*, gives a performance test that measures the effectiveness of vertical strip drains on the time rates of consolidation of compressible soils from construction project sites.

CHAPTER 9
CODES AND PERMITS

9.1 GENERAL

In the concept stages of an urban subsurface drainage project, preferably before the site analysis and system configuration phases, obtaining copies and an understanding of all applicable federal, state, and local codes is important. At the same time, all federal, state, and local permits for the project should be identified, and the requirements and submittal timing of each clearly understood.

9.2 CODES

Federal, state, and local codes that apply to the design, construction, and operation of an urban subsurface drainage system should be considered.

9.3 PERMITS

Federal, state, and local temporary permits that are necessary prior to and during construction of an urban subsurface drainage system project should be secured by the owner, owner's agent, or contractor prior to construction. Any permanent permits, such as the Corps of Engineers' 404 Permit, which must be maintained after construction of the project, should be secured by the owner and/or the owner's agent.

A copy of any temporary and permanent permits secured by the owner or owner's agent should be included as part of the contract documents. Copies of all permits secured by the owner's agent or the contractor should be furnished to the owner.

The contractor is responsible for conforming to the terms and provisions of all permits during construction as stated in the contract documents.

CHAPTER 10
REFERENCES

10.1 CITED REFERENCES

American Association of State Highway Officials (AASHTO), American National Standards Institute (ANSI), American Society of Civil Engineers (ASCE), and ASTM standards are listed in 1.1.

American Concrete Pipe Association. (1998). *Concrete pipe handbook*, ACPA, Irving, TX.

American Concrete Pipe Association. (2004). *Concrete pipe design manual*, ACPA, Irving, TX.

American Railway Engineering and Maintenance-of-Way Association (AREMA). (2010). *Manual for railway engineering*, AREMA, Lanham, MD.

ASCE. (1998). *Urban subsurface drainage*. ASCE Manuals and Reports on Engineering Practice No. 95, ASCE, Reston, VA.

ASCE. (2006). *Standard guidelines for the design of urban stormwater systems*. ASCE/EWRI 45-05, ASCE, Reston, VA.

ASCE. (2007). *Gravity sanitary sewer design and construction*. ASCE Manual on Engineering Practice No. 60, ASCE, Reston, VA.

ASCE. (2009). *Buried Flexible Steel Pipe: Design and Structural Analysis*. ASCE Manual of Practice No. 119, ASCE, Reston, VA.

Chambers, R. E., McGrath, T. J., and Heger, F. J. (1980). *Plastic pipe for subsurface drainage of transportation facilities*. National Cooperative Highway Research Program Report 225, Transportation Research Board, Washington, DC.

Chicago Department of Transportation (CDOT). (2007). *The Chicago green alley handbook*, CDOT, Chicago, IL.

Chow, Ven Te. (1959). *Open channel hydraulics*. McGraw-Hill, New York.

Christopher, B. C., and Zhao, A. (2001). *Design manual for roadway geocomposite underdrain systems*. Tenax Corporation, Baltimore, MD.

Clark, S. E., Steele, K. A., Spicher, J., Siu, C. Y. S., Lalor, M. M., Pitt, R., and Kirby, J. T. (2008). "Roofing materials' contributions to storm-water runoff pollution." *J. Irrig. Drain. Eng.*, 134 (5), 638–645.

Corrugated Polyethylene Pipe Association, Plastics Pipe Institute. (2005). *The complete corrugated polyethylene pipe design manual and installation guide*. Corrugated Polyethylene Pipe Association, Irving, TX.

Davis, S. N., and DeWiest, R. J. M. (1966). *Hydrogeology*, John Wiley & Sons, New York.

Dempsey, B. J. (1988). "Core flow-capacity requirements of geocomposite fin-drain materials used in pavement subdrainage." *Transportation Research Record*, Transportation Research Board, Washington, DC, 1159, 21–29.

Emerson, C. H., and Traver, R. G. (2008). "Multiyear and seasonal variation of infiltration from storm-water best management practices," *J. Irrig. Drain. Eng.*, 134 (5), 598–605.

Federal Aviation Administration, U.S. Department of Transportation. (2006). *Subsurface drainage design*, ACI 150/5230-5C. Washington, DC, Paragraph 7-7.

Federal Highway Administration (FHWA). (1990). *Highway subdrainage design by microcomputer: (DAMP) Drainage analysis and modeling programs*. Publication No. FHWA-IP-90-012. Washington, DC, 84, 88, 101, 108.

FHWA. (1992). *Durable pavement systems, participants' notebook*. Office of Technology Applications and Office of Engineering, Washington, DC.

FHWA. (2002). *User's guide for drainage requirements in pavements – DRIP 2.0 Microcomputer program*. FHWA, Washington, DC, 4–21, 29, 5–8, 14.

Ferguson, B. (1994). *Stormwater infiltration*. Lewis Publishers by CRC Press, Boca, Raton, FL.

Graber, S. D. (2004). "Collection conduits including subsurface drains." *J. Environ. Eng.*, 130 (1), 67–80.

Graber, S. D. (2007a). "Full-flowing collection conduits with nonuniform inflow." *J. Environ. Eng.*, 133 (6), 575–580.

Graber, S. D. (2007b). "Hydraulic considerations in geosynthetic and aggregate subsurface drains." *J. Environ. Eng.*, 133 (9), 869–878.

Harr, M. E. (1962). *Groundwater and Seepage*. McGraw-Hill, New York, 210–226.

Holtz, R.D., Christopher, B.R., and Berg, R.R. (1998). *Geosynthetic Design and Construction Guidelines (Participant Notebook)*, NHI Course No. 13213, FHWA Publication No. HI-95-038. FHWA, Washington, DC.

Koerner, R. M. (2005). *Designing with Geosynthetics,* 5th Ed. Prentice-Hall, Englewood Cliffs, NJ.

Lee, J. G., Heaney, J. P., and Pack, C. A. (2010). "Frequency Methodology for Evaluating Urban and Highway Storm-Water Quality Control Infiltration BMPs," *J. Water Resour. Plann. Manage.*, 136 (2), 237–247.

Marlow, S. (2011a). Personal communication, Technical Sales Manager, American Wick Drain Corporation, January 5, 2011.

Marlow, S. (2011b). Personal Communication, Technical Sales Manager, American Wick Drain Corporation, May 9, 2011.

McEnroe, B. M. (1993). "Maximum Saturated Depth Over Landfill Liner," *J. Environ. Eng.*, 119 (2), 262–270.

McNown, J. S. (1954). "Mechanics of Manifold Flow," *Trans. Am. Soc. Civ. Eng.*, 119, 1103–1142.

National Clay Pipe Institute (NCPI). (2006). *Clay pipe engineering manual*. NCPI, Washington, DC.

National Cooperative Highway Research Program (NCHRP). (2004). *Guide for Mechanistic-Empirical Design of New and Rehabilitated Pavement Structures – Appendix TT: Drainage Requirement in Pavements (DRIP) Microcomputer Program User's Guide*. National Research Council, Washington, DC.

National Corrugated Steel Pipe Association (NCSPA). (2008). *Corrugated steel pipe design manual*. NCSPA, Dallas, TX.

Ohio Department of Transportation (ODOT). (2010). Construction and Material Specifications, Division of Construction Management, Columbus, OH.

Rahman, S., and Watkins, R. K. (2004). "Longitudinal Mechanics of Buried Thermoplastic Pipe: Analysis of PVC Pipes of Various Joint Types," *Proc. Pipeline Division Specialty Conf.*, Aug. 21–24, 2005, Houston, TX, 1101–1116.

Rumer, R. R., Jr., and Drinker, P. A. (1966). "Resistance to Laminar Flow through Porous Media," *J. Hydraul. Div.*, 92 (5), 155–163.

Schneider, L. E., and McCuen, R. H. (2006). "Assessing the Hydrologic Performance of Best Management Practices," *J. Hydrol. Eng.*, 11 (3), 278–281.

Sherard, J. L., Dunnigan, L. P., and Talbot., J. R. (1984a). "Basic Properties of Sand and Gravel Filters," *J. Geotech. Eng.*, 110 (6), 684–700.

Sherard, J. L., Dunnigan, L. P., and Talbot, J. R. (1984b). "Filters for Silts and Clays," *J. Geotech. Eng.*, 110 (6), 701–718.

Smith, G., and Watkins, R. (2004). "The Iowa formula: Its use and misuse when designing flexible pipe," *Proc. Pipeline Division Specialty Cong.*, Aug. 1–4, 2004, San Diego, CA, 1–7.

Somerton, C. W., and Wood, P. (1988). "Effect of walls in modeling flow through porous media," *J. Hydraul. Eng.*, 114 (12), 1431–1448.
Spangler, M. G., and Handy, R. L. (1982). *Soil engineering*, 4th sub Ed. Harpercollins, Glenview, IL.
TeKrony, R. G., Sanders, G. D., and Cummins, B. (2004). "History of drainage in the bureau of Reclamation," *J. Irrig. Drainage Eng.*, 130 (2), 148–153.
Terzaghi, K., Peck, R. B., and Mesri, G. (1996). *Soil mechanics in engineering practice*, 3rd Ed. Wiley Interscience, New York.
U.S. Army Corps of Engineers. (1941). *Investigation of filter requirements for underdrains*. Tech Memo. No. 183-1. USACE, Vicksburg, MS.
USDA, Soil Conservation Service (SCS). (1958). *The structural design of underground conduits*. Technical Release No. 5. USDA, Natural Resources Conservation Service (NRCS), Washington, DC.
USDA. (1969). "Soil Conservation Service, Drainage," Chapter 14. In Engineering Field Manual, USDA, Washington, DC.
USDA. SCS (1971). "Drainage of agricultural land, section 16," *National Engineering Handbook*. USDA, NRCS, Washington, DC.
USDA, SCS. (1980). "Structural design," section 6, *National engineering handbook*. NRCS, Washington, DC.
USDA, SCS. (1991). Subchapter C. Sect. 650, 1428 (b), *Field handbook*. USDA, NRCS, Washington, DC.
U.S. Department of the Interior, Bureau of Reclamation (USBR). (1993). *Drainage manual*. USDI, Denver, CO.
U.S. Occupational Safety and Health Administration, Department of Labor (OSHA). (2002a). Code of Federal Regulations, Title 29, Volume 8, Revised as of July 1, 2002, U.S. Government Printing Office, 29CFR1926.652, pages 377–410, Title 29 – Labor, Chapter XVII, Part 1926 – Safety and Health Regulations for Construction, Subpart P – Excavations, Sec. 1926.652 – Requirements for Protective Systems.
U.S. OSHA. (2002b). Code of Federal Regulations, Title 29, Volume 5, Revised as of July 1, 2002, U.S. Government Printing Office, 29CFR1910.146, pages 467–488, Title 29 – Labor, Chapter XVII, Part 1910 – Occupational Safety and Health Standards, Subpart J – General Environmental Controls, Sec. 1910.146 – Permit-required confined spaces.
Uni-Bell PVC Pipe Association. (2001). *Handbook of PVC pipe: Design and construction*. Uni-Bell PVC Pipe Association, Dallas, TX.
University of New Hampshire Stormwater Center (UNHSC). (2009). *UNHSC Design specification for porous asphalt pavement and infiltration beds*. UNHSC, Durham, NH.
Venkataraman, P., and Rao, P. R. M. (1998). "Darcian, transitional, and turbulent flow through porous media," *J. Hydraul. Eng.*, 124 (8), 840–846 (with errata 125 (4), 437).
Watkins, R. (2004). "Pipe and soil mechanics for buried corrugated HDPE pipe," *Proc. Pipeline Division Specialty Cong.*, Aug. 1–4, 2004, San Diego, CA, 1–10.
Watkins, R., Keil, B., Mielke, R., and Rahman, S. (2010). "Pipe zone bedding: A flexible pipe approach," *Proc. Pipeline Division Specialty Cong.*, Aug. 28-Sept. 1, 2010, Keystone, CO, 426–438.
Ward, J. C. (1964). "Turbulent flow in porous media," *J. Hydraul. Div.*, 90 (5), 1–12.

10.2 GENERAL REFERENCES

American Association of State Highway Officials (AASHTO), American National Standards Institute (ANSI), American Society of Civil Engineers (ASCE), and ASTM standards are listed in 1.1.

Advanced Drainage Systems, Inc. (1984). *Specifier manual*. Advanced Drainage Systems, Columbus, OH.
American Society of Agricultural Engineers. (1982). *Hydrologic modeling of small watersheds*. ASAE, St. Joseph, MI.
ASCE. (1984). "In-Plane composite drains," *Civil engineering*. ASCE, New York, 48–51.
Anderson, B. (1983). *Underground waterproofing*. WEBCO, Stillwater, MN.
Beck, D. E. (1988). "Testing and comparing geocomposite drainage products," *Geotech. Fabrics Report*, Industrial Fabrics Association International, St. Paul, MN, July/August.
Bouwer, H. (1978). *Groundwater hydrology*. McGraw-Hill, New York.
Cedergren, H. R. (1974). *Drainage of highways and airfield pavements*. John Wiley & Sons, New York.
Cedergren, H. R. (1997). *Seepage, drainage, and flow nets*. 3rd Ed. John Wiley & Sons, New York.
Chamber's Technical Dictionary, 3rd Ed. (1967). Macmillan, New York.
Davis, C. V., and Sorensen, K. E. (1986). *Handbook of applied hydraulics*. McGraw-Hill, New York.
Dempsey, B. J. (1988). *Pavement drainage system design*. Prepared for Wisconsin DOT, Madison, WI, February 15–16.
Driscoll, F. G., Ed. (1986). *Ground Water and Wells*, 2nd Ed. Johnson Division UOP, St. Paul, MN.
Engineers Joint Council. (1967). *Thesaurus of engineering and scientific terms*. Engineers Joint Council, New York.
Federal Highway Administration (FHWA). (1979). *Design of highway drainage—The state of the art*, Report No. FHWA-TS-79-225. FHWA, Washington, DC.
FHWA. (1980). *Evaluation of test methods and use criteria for geotechnical fabrics in highway applications*, Report No. FHWA/RD-80-021. FHWA, Washington, DC.
FHWA. (1983). *Design of urban highway drainage*, Report No. FHWA-TS-79-225. FHWA, Washington, DC.
FHWA. (1985). *Highway subdrainage design*, Report No. FHWA-TS-224. FHWA, Washington, DC.
FHWA. (1985). *Hydraulic design of highway culverts*, Hydraulic Design Series No. 5. FHWA, Washington, DC.
Fetter, C. W., Jr. (1980). *Applied hydrogeology*. Charles E. Merrill, Columbus, OH.
Freeze, R. A., and Cherry, J. A. (1979). *Groundwater*. Prentice-Hall, Englewood Cliffs, NJ.
Galloway, P. D. (2008). "Engineer's liability considerations in specifying corrugated high density polyethylene pipe," *J. Prof. Issues in Eng. Ed. Practice*, 134 (1), 87–94.
Geosystems, Inc. (undated). "Vertical drains," *Geonotes—A ground improvements update*. Sterling, VA.
Hancor, Inc. (1993). *Recommended installation practice for Hancor Hi-Q Titelines, heavy duty and heavy duty-AASHTO Pipe*, Hancor, Findlay, OH.
Hannon, J. D., and California DOT. (1980). *Underground disposal of storm water runoff, Design Guidelines Manual*, FHWA-TS-80-218. FHWA, Washington, DC.
Hem, J. D. (1970). *Study and interpretation of the chemical characteristics of natural water*, U.S. Geological Survey Water-Supply Paper 1473, USGS, Washington, DC.
Illinois Department of Transportation. (1993). *Highway standards manual*. Illinois Department of Transporation, Springfield, IL.
Lafayette Farm and Industry. (undated). *Agri-Fabric awareness manual*. Cuba City, WI.
Lohman, S. W., et al. (1972). *Definitions of selected ground water terms-revisions and conceptual refinements*. Geological Survey Water-Supply Paper 1988. USGS, Washington, DC.
Peck, R. B., Hanson, W. E., and Thornburn, T. H. (1974). *Foundation Engineering*. John Wiley & Sons, New York.
Powers, J. P. (2007). *Construction dewatering and groundwater control: New methods and applications*, 3rd Ed. John Wiley & Sons, New York.
Ricketts, J. T., Loftin, M. K., and Merritt, F. S. (2003). *Standard handbook for civil engineers*, 5th ed. McGraw-Hill, New York.
Royster, D. L. (1982). *Landslide remedial measures*. Tennessee Department of Transportation, Nashville, TN.
Sacks, A. (1981). "R_x for basement water problems," *The Family Handyman*, Webb Publishing, St. Paul, MN (currently published by Reader's Digest Association, New York City), September 1981, 36–40.
Sacks, A. M. (1987). "Geosynthetics," *Remodeling*, Hanley Woods, Washington, DC, p. 14, November 1987.
Schuster, R. L., and Kruse, R. J., eds. (1978). *Landslides analysis and control*, Transportation Research Board Special Report 176, Transportation Research Board, Washington, DC.
Schwab, G. O., Revert, R. K., et al. (1981). *Soil and water conservation engineering*, 3rd Ed. John Wiley & Sons, New York.
Sowers, G. F. (1979). *Introductory soil mechanics and foundations: Geotechnical engineering*. Macmillan, New York.
Tensar International Corporation. (2010). "Performance-based specifications for roadways: Discussion on the specification of mechanically stabilized

layers in paved and unpaved applications." White paper. Tensar International, Atlanta, GA.
Todd, D. K. (1980). *Ground water hydrology*. John Wiley & Sons, New York.
USDA. (1988). *Soil conservation service, technical guide*, Section IV, Standard 606, Subsurface Drainage. USDA, Washington, DC.
U.S. Department of the Interior, Bureau of Reclamation (USBR). (1995). *Ground water manual*. USBR, Washington, DC.
U.S. Department of the Interior, USBR. (1997). *Water measurement manual*. USDI, Washington, DC.
Vetch, J. O., and Humphrey, C. R. (1966). *Water and water use terminology*. Thomas Printing and Publishing, New York.

ASCE STANDARD ANSI/ASCE/EWRI 13-13

American Society of Civil Engineers

Standard Guidelines for the Installation of Urban Subsurface Drainage

CONTENTS

FOREWORD

The Board of Direction approved revisions to the ASCE Rules for Standards Committees to govern the writing and maintenance of standards developed by ASCE. All such standards are developed by a consensus standards process managed by the ASCE Codes and Standards Committee (CSC). The consensus process includes balloting by a balanced standards committee and reviewing during a public comment period. All standards are updated or reaffirmed by the same process at intervals between five and 10 years. Requests for formal interpretations shall be processed in accordance with Section 7 of *ASCE Rules for Standards Committees,* which are available at www.asce.org. Errata, addenda, supplements, and interpretations, if any, for these standard guidelines can also be found at www.asce.org.

The *Standard Guidelines for the Installation of Urban Subsurface Drainage* is a companion to the *Standard Guidelines for the Design of Urban Subsurface Drainage* and *Standard Guidelines for the Operation and Maintenance of Urban Subsurface Drainage.* These standard guidelines were developed by the Urban Drainage Standards Committee, which is responsible to the Environmental and Water Resources Institute of the American Society of Civil Engineers.

The provisions of this document are written in permissive language and, as such, offer the user a series of options or instructions but do not prescribe a specific course of action. Significant judgment is left to the user of this document.

These standard guidelines may involve hazardous materials, operations, and equipment. These standard guidelines do not purport to address the safety problems associated with its application. It is the responsibility of whoever uses these standard guidelines to establish appropriate safety and health practices and to determine the applicability of regulatory and nonregulatory limitations.

These standard guidelines have been prepared in accordance with recognized engineering principles and should not be used without the user's competent knowledge for a specific application. The publication of these standard guidelines by ASCE is not intended to warrant that the information contained herein is suitable for any general or specific use, and ASCE takes no position respecting the validity of patent rights. The user is advised that the determination of patent rights or risk of infringement is entirely his or her own responsibility.

ACKNOWLEDGMENTS

The American Society of Civil Engineers (ASCE) acknowledges the work of the Urban Drainage Standards Committee of the Environmental and Water Resources Institute of ASCE (EWRI of ASCE).

This group comprises individuals from many backgrounds, including consulting engineering, research, the construction industry, education, and government. The individuals who serve on the Urban Drainage Standards Committee are

William Curtis Archdeacon, P.E., R.L.S., Past Chair
Kathlie Jeng-Bulloch, Ph.D., P.E., D.WRE, M.ASCE, Chair
Christopher B. Burke, Ph.D., P.E., D.WRE, F.ASCE
Jeffrey S. Glenn, P.E., D.WRE, CFM, F.ASCE
Robert S. Giurato
S. David Graber, P.E., F.ASCE, Vice Chair & Corresp.Editor
Jay M. Herskowitz, P.E., M.ASCE
Conrad G. Keyes, Jr., Ph.D., P.E., L.S., D.WRE, Dist.M.ASCE
James H. Lenhart, Jr., P.E., D.WRE, M.ASCE
Lawrence M. Magura, P.E., M.ASCE, Secretary
Garvin J. Pederson, P.E., M.ASCE
Anthony N. Tafuri, P.E., D.WRE, M.ASCE
Kenneth E. Waite, P.E., M.ASCE
William J. Weaver, P.E., M.ASCE
Donald E. Woodward, P.E., M.ASCE
Bin Zhang, P.E.

The corresponding editor recognizes the following committee members who were particularly helpful in updating the standard guidelines: Jeffrey S. Glenn, Jay M. Herskowitz, Conrad G. Keyes, Jr., James H. Lenhart, and William J. Weaver.

CHAPTER 1
SCOPE

The intent of these standard guidelines is to present installation and construction guidance for urban subsurface drainage systems. It updates ASCE/EWRI 13-05 with material developed within the past five years. The collection and conveyance of subsurface drainage waters are within the purview of these standard guidelines for applications such as airports, roads, and other commercial transportation systems and for industrial, residential, and recreational areas. Incidental surface water is considered.

These standard guidelines do not address agricultural drainage, landfills, recharge systems, detention ponds, conventional storm sewer systems, or the use of injection systems.

Customary units and standard international (SI) units are used throughout this document.

1.1 APPLICABLE STANDARDS

Standards, including those applicable to these guidelines, are given in 1.1 of ANSI/ASCE/EWRI 12.

CHAPTER 2
DEFINITIONS

2.1 GENERAL

This chapter defines specific terms for use in these guidelines. References included in Section 10 of ANSI/ASCE/EWRI 12 may also be helpful in understanding the terms of these guidelines.

2.2 TERMS

Bedding—Granular material placed around subsurface drains to provide structural support for the drain.

Drain envelope—Generic name for materials placed on or around a drainage product, regardless of whether used for mechanical support (bedding), hydraulic purposes (hydraulic envelope), or stabilizing surrounding soil material (filter envelope).

Filter envelope—Permeable material placed around a drainage product to stabilize the structure of the surrounding soil material. A filter envelope may initially allow some fines and colloidal material to pass through it and into the drain.

Geocomposite—Geosynthetic materials for collecting and transporting water while maintaining soil stability.

Geomembrane—Sheet materials intended to form an impervious barrier.

Geosynthetic—Synthetic material or structure used as an integral part of a project, structure, or system. Within this category are subsurface drainage and water control materials such as geomembranes, geotextiles, and geocomposites.

Geotextile—Woven or nonwoven engineering fabric sheet material intended to allow the passage of water (but not fines and without collecting fines at the soil–textile interface).

Grade—May refer to either (1) the slope of the drain in vertical units/horizontal units, or (2) the specified vertical location of the drain, depending on the context in which it is used.

Hydraulic envelope—Permeable material placed around a drainage product to improve flow conditions in the area immediately adjacent to the drain.

OSHA—Occupational Safety and Health Administration, the federal agency responsible for safety and health concerns on construction job sites.

Subsurface water—All water beneath the ground or pavement surface. Sometimes referred to as groundwater.

CHAPTER 3
CONTRACT DOCUMENTS

3.1 GENERAL

Plans and specifications typically make up the construction contract documents. Other documents become part of the contract by reference.

3.2 PLANS

Plans are drawings, most prepared to a scale, showing the proposed subsurface drainage system and known surface and subsurface features that may affect the new installation and construction. The plans typically show type, size, material, grade, and location of the subsurface drainage system.

3.3 SPECIFICATIONS

Specifications are written text and/or details that provide specifics of the construction. They typically detail or reference all appropriate standards required for the product and project. Construction methods may be specified on a case-by-case basis.

3.4 OTHER

Many other publications are made part of the contract documents by reference only such as government requirements, permits, reports, or trade and industry specifications. Some of these reports include soil borings, past construction observations, and studies.

CHAPTER 4
SITE INVESTIGATION

4.1 GENERAL

Examining the plans and specifications and making a personal examination of the site and its surroundings prior to construction are necessary. This investigation should include reviews of both surface features and subsurface features.

4.2 SURFACE FEATURES

The surface features of the site should be located through a topographic survey and shown on the plans. The plans should be compared to existing field conditions to determine whether any differences exist between the topographic survey and present conditions. Discrepancies are to be brought to the attention of the project engineer.

4.3 SUBSURFACE FEATURES

Subsurface features principally consist of utilities, water table level, and geologic conditions. All subsurface features affecting the work should be determined and shown on the plans. If called for in the plans, the contractor shall undertake a program of field surveys and test pits to verify subsurface conditions at specific locations identified in the plans. The contractor shall immediately notify the project engineer if changed conditions are found and shall not commence construction of the work in areas where discrepancies are noted until the project engineer has issued written clarification.

4.3.1 Utilities. The location and size of sanitary sewers, drains, culverts, gas lines, water mains, electric lines, telephone conduits, and other underground utilities and structures should be shown on the plans. This information should be obtained from both field surveys and other available records. Utility-locating organizations are frequently available in urban areas to mark the position of lines. The level of confidence for utility location and depth may require special provisions to be placed in the contract documents, indicating areas where contractors must physically locate existing utilities before constructing drain lines.

4.3.2 Geologic Conditions. All appropriate and available geologic conditions should be shown on the plans. An assessment should be made with respect to rock and groundwater conditions.

CHAPTER 5
INSTALLATION

5.1 GENERAL

Prior to construction, all documents, including plans and specifications, subsurface information, standard details, product shop drawings, and special provisions should be reviewed and any questions resolved.

5.2 SAFETY

The contractor is responsible for construction site safety. U.S. federal regulations covering safety for all types of construction are published in the Safety and Health Regulations for Construction under the Department of Labor, Occupational Safety and Health Administration (OSHA). Many states, municipalities, and other local agencies have established codes of safe practice regarding construction. These regulations apply to all types of construction, including alteration and repair work. All personnel associated with the construction should be familiar with the requirements applicable to subsurface drainage system projects, especially in regard to safe trenching procedures.

5.3 SOIL EROSION AND SEDIMENT CONTROL

Erosion and sediment control at the site should be in accordance with federal, state, municipal, and local agency regulations and as otherwise established by the contract plans and specifications.

5.4 SITE PREPARATION

Site preparation should be in accordance with the contract plans and specifications and may include topsoil stripping, clearing and grubbing, pavement and sidewalk removal, rough grading, protection or relocation of existing natural drainage, removal of unsuitable soil material, construction of access roads and detours, and protection or relocation of existing structures and utilities.

5.5 MATERIALS RECEIVING, HANDLING, AND STORAGE

The contractor is responsible for receiving, proper handling, and storage of all construction materials for the project. Materials damaged in shipment or at the site that cannot be repaired should be marked or tagged and removed from the site. All materials should be unloaded and handled with reasonable care. Stockpiling of materials should be as near as possible to where they will be installed, consistent with municipal requirements, safety, and environmental considerations. All materials should be stored as recommended by the manufacturer. All materials should be stockpiled in a safe manner.

5.6 LINE AND GRADE

The contractor is responsible for maintaining all line and grade, monuments, control points, and stakes set by the licensed surveyor until the project is completed and accepted.

For all subsurface drainage systems, line and grade should be in accordance with the contract plans and specifications.

All subsurface drainage systems should be installed true to line and grade in accordance with the contract plans and specifications. Adjustments to correct departures from specified line and grade shall not exceed those permitted by contract documents or manufacturers' recommendations, whichever are more restrictive, provided that such corrections never result in a reversal of the slope in the drainage system. Moreover, realignments must never result in buckling or other deformations of flexible drainage materials, such as flexible pipes and geocomposites, that damage or reduce the flow capacity of the system. The return to specified line and grade should be made by adding or removing bedding material, and the use of wedges or blocks is unacceptable.

The maximum allowable departure from horizontal alignment should be specified by the contract documents. Departure distances must be determined by measurement along common elements of the planned versus actual installation, such as centerline-to-centerline measurements.

The relatively complex hydraulic characteristics of geocomposite drainage systems require that strict controls be placed on departures from specified vertical alignment (grade). When departures occur, the return to specified grade should be at a rate not exceeding 15% of the specified grade or at a rate established by the specifications.

5.7 EXCAVATION

Excavation should be in accordance with contract plans and specifications and may include trenching, backfilling, embankment construction, soil stabilization, and control of groundwater and surface drainage. Adequate knowledge of subsurface conditions is required for all types of excavation. Additional exploration and analysis are recommended if the subsurface information on the plans is insufficient. The contractor should notify the owner if archaeological items are encountered during construction.

5.7.1 Excavation Limits. Excavation, installation, and backfill operations should be performed in a timely manner to reduce open trench time. The length of open trench should comply with limits established by the federal safety and health administration (e.g., OSHA) or applicable state and local regulations. Specified trench width requirements for flexible conduits should also be maintained to ensure proper deflection control. Trench depth and width should be in accordance with the contract plans and specifications. If the trench width becomes greater than specified, the contractor should contact the project engineer for a reevaluation of the required pipe strength, bedding materials and methods, and backfilling procedures to be used.

5.7.2 Handling of Excavated Material. Excavated material to be used as backfill should be stockpiled in accordance with the contract plans and specifications and applicable safety regulations a safe distance back from the edge of the trench. Generally, if trench walls are unsupported, the minimum distance from the trench side to the excavated material should be either 3 ft (1 m) or one-half of the trench depth, whichever is greater. If the trench walls are supported, the usual minimum distance from the trench side to the excavated material should be 3 ft (1 m).

These general recommendations should not supersede job-specific requirements in the presence of unstable soils and/or the potential for accumulation of water in the trenches.

For trench installations where drainage material and backfill are placed simultaneously and where personnel are not allowed in the trench, soil placement may be unrestricted. Unused excavated materials should be disposed of in accordance with contract documents.

5.7.3 Sheathing and Shoring. OSHA and many states, municipalities, and other local agencies have established codes of safe practice regarding support requirements for trench excavation. Sheathing and shoring must be adequate to prevent cave-in of the trench walls or subsidence of areas adjacent to the trench and to prevent sloughing of the base of the excavation. Sheathing and shoring should not extend into the soil envelope zone of the pipe or geocomposite drainage system. Any sheathing placed below the top of the drainage product of flexible pipes or geocomposite drainage materials or the springline of rigid pipes should remain in place after backfilling. Movement of shoring following backfill placement may reduce the structural integrity of the surrounding embedment material.

The contractor is responsible for adequacy of any required sheathing and shoring. The strength of support systems should be based on the principles of geotechnical and structural engineering as applicable to the materials encountered. Refer to Appendix A for recommended use of trench boxes.

5.7.4 Dewatering. When necessary, all excavations should be dewatered prior to and during installation and backfilling of the subsurface drain system. The contractor is responsible for dewatering operations and should ensure that foundation and bedding materials are not being removed through the dewatering system and that property damage does not result.

The discharge from dewatering systems should be released in such a way that it does not harm existing structures or the environment and that it does not, by reinfiltration, put an additional load on the dewatering system.

5.8 FOUNDATION PREPARATION

Bedding for subsurface drainage systems should be as specified and completed to design line and grade. The intrusion of foreign material into any portion of the drainage system due to construction and weather events should be prevented until the system is adequately protected by backfill.

5.9 PLACEMENT OF DRAINAGE MATERIALS

The pipe materials covered are pipe, geocomposites, and other. Installation recommendations are available from standards, handbooks, other publications, and manufacturers. Examples from the references provided in ASCE/EWRI 12 are NCSPA (2008) for corrugated steel pipe; Corrugated Polyethylene Pipe Association (2005), ASTM D2321, and Rahman and Watkins (2004) for thermoplastic pipe; and ASCE 15-98 and American Concrete Pipe Association (1998) for precast concrete pipe.

5.9.1 Pipe. A bedding material should be placed on the foundation, the pipe laid and connected, and backfill placed, all in accordance with the contract plans and specifications.

5.9.2 Geocomposite Drainage Materials. Prefabricated geocomposite subsurface collector drains may be placed in trenches by hand or machine. Installation techniques must not cause damage to the interior core and geotextile overwrap, including any factory-made seams and connections. Joints should be made using connectors recommended by the manufacturer or in accordance with the contract plans and specifications. All joints should be made soil tight using tape, glue, or other sealing procedures recommended by the manufacturer or in accordance with the contract plans and specifications. In all cases, such sealing procedures should provide ensured long-term resistance to degradation in wet subsurface environments.

5.9.3 Other Drainage Materials. Geomembranes, geotextiles, aggregates, and pump and lift stations should be installed in accordance with the contract plans and specifications.

5.10 BACKFILL

Backfill material should be placed and compacted in accordance with the contract plans and specifications. Backfill should not be dumped or dropped directly on any portion of the drainage system. Heavy equipment operations should be controlled so as not to damage any portion of the drainage system. Backfill material should be placed in layers in accordance with contract documents and should be compacted.

If infiltration and groundwater recharge are an objective, direction regarding protecting the subgrade from excessive compaction may need to be provided along with an understanding of subgrade infiltration rates. For porous asphalt pavement and infiltration beds, the existing native subgrade material under all bed areas should not be compacted or subject to excessive construction equipment traffic prior to geotextile and stone bed placement (UNHSC 2009). Compaction is acceptable if an impermeable liner is used at the base of the porous asphalt system and infiltration is not an objective.

5.11 SITE RESTORATION

Restoration of grass, shrubs, and other plantings should be performed in accordance with contract documents. Until revegetation is complete, adequate protection against erosion and runoff is necessary and should be in accordance with the contract documents and governing regulations. All revegetation and tree repair should be in accordance with accepted horticultural practice.

When replacing permanent pavement, the subgrade must be restored and compacted until smooth and to specified density. Thickness and type of pavement shall be as established by contract documents.

CHAPTER 6
INSPECTION

6.1 GENERAL

The duty of the inspector is to examine the materials furnished and the work performed to verify full compliance with the contract documents. A dedicated, qualified inspector should be on site to observe all phases of the site preparation, materials receiving and handling, installation, and site restoration. Observations of materials, workmanship, and, where specified in the contract documents, methods and means of performing construction are required to determine compliance with contract documents.

The inspector should have unrestricted access to all areas where the preparation of the materials and parts of work to be done are carried out and conducted. The contractor shall provide access to all facilities and assistance required to perform the inspection.

6.2 INSPECTION OF MATERIALS

All construction materials must be carefully and thoroughly inspected prior to and during placement. Inspection should be an ongoing process, because materials that are satisfactory when first arriving on site can be damaged during handling, storage, and installation. No material of any kind should be used prior to inspection and formal approval.

Project specifications and other product-specified information should form the basis for determining the suitability of all materials. Should any doubt concerning suitability arise, the product manufacturer should be consulted.

Shipments of select fill materials and drainage products should be accompanied by certified test reports. If such data are missing, laboratory tests should be used to confirm appropriate properties. All drainage products or drainage system components should be measured to check size, shape, and fit.

All materials must be inspected to ensure that they are free of foreign deposits, defects, and damage. Cleaning and removal of foreign matter may be acceptable, provided there is complete assurance that the construction material or product is unharmed and in "like new" condition.

Damaged products or components should be immediately removed from the site. Repairs may be performed on damaged goods following inspection and approval of the inspector and after consultation with the manufacturer. The return of any previously rejected materials, products, or components to the site is acceptable only after reinspection and approval following rework. Any material or workmanship found at any time during the construction cycle not in accordance with project specifications, for any reason, shall immediately be remedied.

6.2.1 Prefabricated and/or Premanufactured Components. Prior to installation, all prefabricated and/or premanufactured components shall be inspected to establish conformity with the project specifications and to check for damage and the presence of foreign matter. The manufacturer's certificate of compliance and product drawings should confirm compliance with the contract documents.

6.2.2 Bedding, Backfill, and Envelope Materials. All materials for use in bedding, backfilling, and envelope materials, or as otherwise used in subsurface drainage systems, shall be checked for conformance to project specifications. The supplier's certificate of compliance may form the basis of the inspection.

Geosynthetic products should be inspected for damage and conformity to the project specifications and drawings. The manufacturer's materials certificate should be the basic inspection document. Any deviations from the contract specifications should be immediately referred to the project engineer.

6.2.3 Storage of Materials. Storage of materials should be managed by the contractor to avoid impairing the usability and quality of on-site materials. Observance of any special handling methods required shall be verified and recorded. Storage or special protection required by the contract documents for certain items should also be verified. The inspector is responsible for monitoring the contractor's observance of these requirements.

6.3 INSPECTION OF EQUIPMENT

The contractor should verify that the equipment provided is appropriate in size for performance of the contract, is in good operating condition, and meets safety requirements.

6.4 INSPECTION OF CONSTRUCTION

6.4.1 General. The sequence of construction operations is an important consideration in projects that require construction in a particular order. Sequencing may also be required to allow the existing facilities to remain in operation during the construction phase. The contract documents normally allow as much flexibility as possible in sequencing operations and may have no requirements other than an overall completion date.

The inspector's function is not to supervise or direct the manner in which the work is performed. The inspector should follow each stage of construction so that any construction errors can be resolved during construction rather than after. The inspector should closely monitor acceptance testing for the correct test procedure. Inspection activities associated with each construction stage are summarized in Table 6-1.

6.4.2. Construction Layout. Survey controls should be established as referenced on the contract drawings. The contractor is responsible for staking and building the project from the controls. All necessary auxiliary staking must be in place prior to construction. If an error in auxiliary staking is observed, detected, or suspected, the error shall be promptly called to the contractor's attention.

Table 6-1. Inspection Activities Associated with Construction

Construction Activity	Inspection Activity
1. Site preparation	Monitor clearing and grubbing and disposition of topsoil
2. Subgrade/foundation preparation	Observation and verification of grades Verification of suitability of subgrade
3. Receiving/storing materials	Verification of type and condition of materials received at site and storage procedures
4. Trench excavation	Observation and verification of width, locations, lines, and grades
5. Installation of subsurface drainage systems	Verification of lines and grades and testing where necessary
6. Initial backfilling	Verification of proper filter layer placement Verification of use of proper placement and compaction technique Verification of specific gradation, thickness, and densities
7. Final backfilling	Verification of proper backfill material placement procedure and compaction procedure Test verification of specified density and moisture content
8. Site restoration	Verification of final site conditions as specified

The inspector is responsible for ascertaining that the survey control points are in place as referenced. If any evidence exists that the control points have been disturbed, the inspector shall notify the licensed engineer, who will arrange to have the points checked by the licensed surveyor who originally set the points or take other appropriate action.

6.4.3 Excavation and Dewatering. The inspector should confirm that all excavations and dewatering activities are performed in accordance with the contract documents and that these activities will allow construction to be completed according to plan. Proper disposal of water is the responsibility of the contractor.

Trenches shall be excavated to depths and widths as specified for correct backfill and/or envelope placement and compaction. Water standing or flowing into the trench should be removed until backfill and envelope materials are placed. When possible, trenching should occur across slope and not downslope to minimize soil loss during rain.

Prior to placement of drainage products and envelope materials, all finished excavations should be inspected to ensure the absence of unsuitable materials.

6.4.4 Installation. Where the contract documents reference installation in accordance with the manufacturer's directions, such directions shall be provided for use in verifying that subsequent installation is performed in accordance with them.

6.4.5 Backfilling. The drain should be inspected for proper elevation, grade, alignment, and joint spacing; collapsed, broken, or cracked pipe; and thickness of aggregate envelope before backfilling. Backfill placement should be in accordance with 5.10.

6.4.6 Televising. The underdrain should be inspected by means of closed-circuit television or other acceptable camera systems where appropriate. Permanent videotape or film should be furnished in accordance with the contract documents.

6.4.7 Testing. Few definitive tests can be performed on installed subsurface drains that give measurable indications of the functional effectiveness of the installation.

The specifications may require specific field tests to be performed. As appropriate, samples shall be furnished by the contractor or representative samples will be taken from delivered materials. The number of samples shall be sufficient to satisfy all testing requirements. Control testing shall be performed in the field or at such other locations as required. The overall inspector, together with the project engineer, will review the testing requirements and determine the overall testing program including all necessary testing facilities and record forms will be specified.

Complete records of the test and results shall be retained. Specimens shall be retained if they are important to prove results of the specified tests. Before being accepted as completed, each drain should be tested for obstructions and any deflections or deformations.

6.4.8 Safety. All safety rules established by governmental agencies shall be strictly followed. Observations should be performed in a manner that will not unreasonably impede or obstruct the contractor's operations.

6.4.9 Suitability and/or Conformance. Methods and means of construction are left to the option of the contractor on most items to allow flexibility, but they may be specified on items where methods and means are critical to obtaining a final desired product. Whereas means and methods may be specified in full detail, minimum elements are frequently identified. Where methods and means are specified in the contract documents, verification is required for compliance.

6.5 ACCEPTANCE OF CONSTRUCTION

Acceptance of construction normally covers the entire job and is not done on an incremental basis. In normal day-to-day operations, field personnel may verbally acknowledge apparent compliance with the contract documents. However, such acknowledgment should not constitute acceptance of part or all of the construction.

6.6 RECORDING OBSERVATIONS

Work performed by the contractor on a shift basis should be recorded by the inspector to provide a detailed record of the progress. All observations of noncompliance with the contract documents shall be recorded in the daily report. The report should cover any verbal statements made to and by the contractor concerning the noncompliance. Photographs or videos should be taken when they assist in describing the noncompliance.

On matters not immediately corrected, the inspector should give the contractor a separate written notice of noncompliance within 24 hours. The notice should state specifically why the work does not meet the requirements of the contract documents.

6.7 RECORD DRAWINGS

Record details of construction should be incorporated into a final revision of the construction drawings to represent the most reliable record for future use.

During construction, the contractor and/or inspector should measure, reference, and record the locations of all inlets, outlets, and stubs for future connections and other buried facilities. All construction changes from the original plans should be recorded. Contract drawings should be revised to indicate the field information after the project is completed. A notation, such as "Revised According to Field Construction Records" or "Record Drawing," should be made on each sheet along with the contractor's and/or inspector's name, date, and company name. Records of such plans should become a part of the owner's permanent records.

For subsurface drainage systems, the following minimum information should be included on the record drawing:

- Size and type of all drains on plan and profile sheets;
- Station and pipe invert elevation of all tees, wyes, cleanouts, manholes, and outfalls; and
- Manholes and other critical points referenced to established survey points.

CHAPTER 7
REFERENCES

Standards, including those applicable to these guidelines, are given in 1.1 of ANSI/ASCE/EWRI 12. Other references, including those applicable to these guidelines, are given in Chapter 10 of ANSI/ASCE/EWRI 12.

APPENDIX A
RECOMMENDED USE OF TRENCH BOXES

A1.0 INTRODUCTION

Trench boxes provide a safer work area to install pipe in deep trenches or in soils that have insufficient stability. Use of a trench box may also be required by the specifications for reasons other than safety. Whereas these guidelines will work for most cohesive and noncohesive native soils, highly unusual soil conditions may require further investigation.

A2.0 GENERAL CONSIDERATIONS

Some installations may not require trench boxes if the trench sidewall can be sloped. The licensed engineer should provide specific guidance on acceptable slopes, but in no case should the trench wall slope be greater than the angle of repose of the native soil.

The length of the trench box should be suitable for the pipe length.

A3.0 SUBTRENCH CONSTRUCTION

The most effective way to maintain a sound trench system is to provide a subtrench within which to place the pipe and backfill. Backfill and compact according to the design specifications within the subtrench. The trench box can be pulled along the top edge of the subtrench (Fig. A3-1) without affecting the pipe or the backfill.

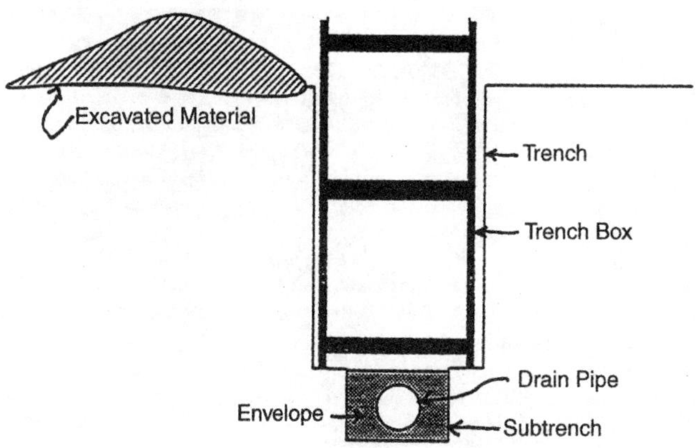

FIGURE A3-1. Subtrench construction

Subtrench construction also makes using a geotextile around the backfill easier if it is required by the project specifications. Line the subtrench with the geotextile, place the pipe and backfill, and wrap the geotextile over the top of the pipe, and backfill system.

A4.0 REGULAR TRENCH CONSTRUCTION

In construction not involving a subtrench situation, dragging a trench box subjects the pipe to stretching or separated joint. The box should be lifted vertically until it is above the pipe and reset into its new position. If dragging a trench box through a trench is necessary, do not lower the box more than one-fourth of the nominal diameter below the crown (top) of the pipe. This allows the backfill material to flow out of the bottom of the box around the pipe so that backfill disturbance is minimized.

An alternative for flexible stormwater system when the box will be dragged is to use a well-graded granular backfill material two diameters on either side of the pipe and compact it to a minimum of 90% standard Proctor density before moving the box. Immediately fill the area between the pipe and backfill structure and the trench wall with a granular material.

If the project requires a geotextile around the backfill, use a well-graded granular backfill material and compact it to at least 90% standard Proctor density. Do not drag the box; instead, lift it vertically. After the trench box is removed, immediately fill the area between the pipe and backfill structure and the trench wall with a granular material and compact according to project requirements. The geotextile manufacturer may be able to provide additional information regarding the suitability of specific geotextiles for use with trench boxes.

A5.0 SUMMARY

While trench boxes increase worker safety in difficult site conditions, precautions are required to ensure a structurally sound pipe and backfill system.

Construction of a subtrench is the most effective means of maintaining a sound system. When a regular trench is used, techniques such as lifting the box, keeping the box about three-fourths the nominal diameter up from the trench bottom, and providing a wide granular backfill envelope will help ensure quality construction.

American Society of Civil Engineers

Standard Guidelines for the Operation and Maintenance of Urban Subsurface Drainage

CONTENTS

FOREWORD

The Board of Direction approved revisions to the ASCE Rules for Standards Committees to govern the writing and maintenance of standards developed by ASCE. All such standards are developed by a consensus standards process managed by the ASCE Codes and Standards Committee (CSC). The consensus process includes balloting by a balanced standards committee and reviewing during a public comment period. All standards are updated or reaffirmed by the same process at intervals between five and 10 years. Requests for formal interpretations shall be processed in accordance with Section 7 of *ASCE Rules for Standards Committees,* which are available at www.asce.org. Errata, addenda, supplements, and interpretations, if any, for these standard guidelines can also be found at www.asce.org.

The *Standard Guidelines for the Operation and Maintenance of Urban Subsurface Drainage* is a companion to the *Standard Guidelines for the Design of Urban Subsurface Drainage* and *Standard Guidelines for the Installation of Urban Subsurface Drainage.* These standard guidelines were developed by the Urban Drainage Standards Committee, which is responsible to the Environmental and Water Resources Institute of the American Society of Civil Engineers.

The provisions of this document are written in permissive language and, as such, offer the user a series of options or instructions but do not prescribe a specific course of action. Significant judgment is left to the user of this document.

These standard guidelines may involve hazardous materials, operations, and equipment. These standard guidelines do not purport to address the safety problems associated with its application. It is the responsibility of whoever uses these standard guidelines to establish appropriate safety and health practices and to determine the applicability of regulatory and nonregulatory limitations.

These standard guidelines have been prepared in accordance with recognized engineering principles and should not be used without the user's competent knowledge for a specific application. The publication of these standard guidelines by ASCE is not intended to warrant that the information contained therein is suitable for any general or specific use, and ASCE takes no position respecting the validity of patent rights. The user is advised that the determination of patent rights or risk of infringement is entirely his or her own responsibility.

ACKNOWLEDGMENTS

The American Society of Civil Engineers (ASCE) acknowledges the work of the Urban Drainage Standards Committee of the Environmental and Water Resources Institute of ASCE (EWRI of ASCE).

This group comprises individuals from many backgrounds, including consulting engineering, research, the construction industry, education, and government. The individuals who serve on the Urban Drainage Standards Committee are

William Curtis Archdeacon, P.E., R.L.S., Past Chair
Kathlie Jeng-Bulloch, Ph.D., P.E., D.WRE, M.ASCE, Chair
Christopher B. Burke, Ph.D., P.E., D.WRE, F.ASCE
Jeffrey S. Glenn, P.E., D.WRE, CFM, F.ASCE
Robert S. Giurato
S. David Graber, P.E., F.ASCE, Vice Chair & Corresp.Editor
Jay M. Herskowitz, P.E., M.ASCE
Conrad G. Keyes, Jr., Ph.D., P.E., L.S., D.WRE, Dist.M.ASCE
James H. Lenhart, Jr., P.E., D.WRE, M.ASCE
Lawrence M. Magura, P.E., M.ASCE, Secretary
Garvin J. Pederson, P.E., M.ASCE
Anthony N. Tafuri, P.E., D.WRE, M.ASCE
Kenneth E. Waite, P.E., M.ASCE
William J. Weaver, P.E., M.ASCE
Donald E. Woodward, P.E., M.ASCE
Bin Zhang, P.E.

The corresponding editor recognizes the following committee members who were particularly helpful in updating the standard guidelines: Jeffrey S. Glenn, Jay M. Herskowitz, Conrad G. Keyes, Jr., James H. Lenhart, and William J. Weaver.

CHAPTER 1
SCOPE

The intent of these standard guidelines is to present operation and maintenance guidance for urban subsurface drainage systems. It updates ASCE/EWRI 14-05 with material developed within the past five years. The collection and conveyance of subsurface drainage waters are within the purview of these standard guidelines for applications such as airports, roads, and other commercial transportation systems and for industrial, residential, and recreational areas. Incidental surface water is considered.

These standard guidelines do not address agricultural drainage, landfills, recharge systems, detention ponds, conventional storm sewer systems, or the use of injection systems.

Customary units and standard international (SI) units are used throughout this document.

1.1 APPLICABLE STANDARDS

Standards, including those applicable to these guidelines, are given in 1.1 of ANSI/ASCE/EWRI 12.

CHAPTER 2
DEFINITIONS

2.1 GENERAL

This chapter defines specific terms for use in these guidelines. The reference documents listed in Section 10 of ANSI/ASCE/EWRI 12 may also be helpful in understanding terms in these standard guidelines.

2.2 TERMS

Aquifer—Water-bearing stratum of permeable rock, sand, or gravel.

Bedding—Support for pipe during the construction process.

Drain envelope—Generic name for materials placed on or around a drainage product, regardless of whether used for mechanical support (bedding), hydraulic purposes (hydraulic envelope), or stabilizing surrounding soil material (filter envelope).

Filter envelope—Permeable material placed around a drainage product to stabilize the structure of the surrounding soil material. A filter envelope may initially allow some fines and colloidal material to pass through it and into the drain.

Geocomposite—Geosynthetic materials for collecting and transporting water while maintaining soil stability.

Geomembrane—Sheet materials intended to form an impervious barrier.

Geosynthetic—Synthetic material or structure used as an integral part of a project, structure, or system. Within this category are subsurface drainage and water control materials such as geomembranes, geotextiles, and geocomposites.

Geotextile—Woven or nonwoven engineering fabric sheet material intended to allow the passage of water (but not fines and without collecting fines at the soil-textile interface).

Hydraulic envelope—Permeable material placed around a drainage product to improve flow conditions in the area immediately adjacent to the drain.

Inspection—Critical examination to determine conformance to applicable quality standards or specifications.

Iron ochre—Red or yellow gelatinous deposit formed by a combination of soluble iron deposits and bacteria.

Jetting—Method used to clean pipes involving high-pressure water.

OSHA—Occupational Safety and Health Administration, United States Department of Labor, the United States federal agency responsible for safety and health concerns on construction job sites.

Rodding—Method used to clean pipes involving mechanical means.

Record drawing—Drawings prepared during or after construction showing the final measurements of construction, including any deviations from the design drawings and certain other field observations such as tie-in locations.

Subsurface water—All water beneath the ground or pavement surface, sometimes referred to as groundwater.

USEPA—United States Environmental Protection Agency.

CHAPTER 3
OPERATION AND MAINTENANCE PLAN

3.1 GENERAL

This section provides a general outline of what should be included in an operation and maintenance plan. Modification of this plan may be necessary to more accurately reflect the specific subsurface drainage system under consideration.

This is intended as general standard guidelines for preparing procedures and timetables related to routine operation and maintenance of subsurface drainage systems. Technical personnel must be familiar with the basic concepts of subsurface drainage facilities.

Operation and maintenance instruction materials submitted by manufacturers supplying equipment for urban subsurface system components should be retained and incorporated into the procedures document. The manual should explain the general operational relationships among the various system components of the facility and include any manufacturers' instructions or recommendations. Adherence to these procedures is essential to retain the operating capacity of a facility throughout its expected service life.

3.2 SCOPE

An operations and maintenance plan is intended to provide general guidance for the topics listed. However, this list of subjects may not cover all aspects of any particular subsurface drainage system.

- Normal operating procedures to be followed during the majority of the facility's service life,
- Special operating procedures to be followed when normal procedures cannot be followed,
- Routine maintenance procedures specifically described according to a definite schedule,
- Safety procedures for all operating personnel, and
- Emergency procedures for conditions involving serious service disruptions.

3.3 RESPONSIBILITIES

The owner has the responsibility for the operation and maintenance of the subsurface drainage system. Generally, these areas of responsibility include

- Providing adequate funds for supplies and personnel, operation and maintenance of equipment, and any necessary system expansion;
- Selecting either trained personnel or providing training and education for qualified individuals;
- Developing and implementing a comprehensive program that includes a complete reporting and records retention system; and
- Coordinating with other departments for maintenance and operation to ensure an integrated system is working.

3.4 DESIGN CRITERIA

The procedures manual should detail the performance criteria used in the design. All personnel should be aware of these criteria so that deviations can be recognized.

3.5 NORMAL OPERATING PROCEDURES

Normal operation is the design condition expected of the system. When all components are functioning as designed, the only requirements of the operating personnel are routine checks and scheduled maintenance.

3.6 ABNORMAL OPERATING PROCEDURES

3.6.1 General. There are several types of abnormal operating procedures to consider.

3.6.2 Line Blockage. Subsurface drains are often designed for minimal flow. As a result, subsurface drains are susceptible to blockage. If a pipeline becomes blocked, rodding or jetting may be required to clear it. During cleaning operations, a careful watch should be maintained at the downstream manhole for an indication of the cause of the blockage.

Cleanout assemblies may be located at periodic intervals along the pipeline as shown on the "as-built/ record drawings." If a conveyance main becomes blocked, the nearest of these cleanouts can be located, accessed, and cleaning equipment inserted.

Hydraulic flushing may be attempted to clean any pipeline. Special care should be taken to avoid damage caused by hydraulic surging.

3.6.3 Other System Components. Other components should be operated according to manufacturers' recommendations. In the event of reduced performance of system components, correction procedures in compliance with manufacturers' recommendations should be used.

3.7 MAINTENANCE PROCEDURES

3.7.1 General. Maintenance can be broadly classified as either corrective or preventive.

3.7.2 Corrective Maintenance. Corrective maintenance involves the repair of equipment after breakdown. The operation and maintenance plan should include all necessary equipment manuals, diagrams, and instructions for satisfactory operation.

3.7.3 Preventive Maintenance. Equipment breakdown often relates to a failure of preventive maintenance. As the term implies, preventive maintenance is intended to prevent disruptive breakdowns. Because many components that require different preventive maintenance actions at different time intervals may be involved in the system, preventive maintenance is best performed on a scheduled basis from a checklist. Manufacturers' recommendations should be followed during preventive maintenance.

3.7.4 Checklists. Checklists should be maintained for each component of the urban subsurface system with recommended schedules for inspection clearly stated. Checklists should include each aspect of the inspection such as damage to structure, evidence of restricted capacity, and manufacturers' recommended maintenance. Once all inspections outlined on the form have been completed, the forms should be replaced and the completed forms filed in the owner's records.

3.7.5 Annual System Inspection. A general overview inspection of the entire system should be performed at least annually in addition to the more intensive inspections performed as discussed in Chapter 5 of these standard guidelines. Local conditions may make more frequent inspections of certain components necessary. These may be adjusted as experience is gained in the operation of the system but in no case should the annual inspection be extended.

3.8 SAFETY

3.8.1 General. All personnel are responsible for keeping areas safe and clean. Guards should be in place on operating equipment, and all areas should be properly lit. All enclosed space should be adequately ventilated prior to personnel entering. All personnel should be sure they understand the following:

- Location of all safety equipment;
- Use of safety equipment and devices;
- OSHA, USEPA, and local safety rules; and
- Potential for "danger."

3.8.2 Mechanical. When working on pumps, suction and discharge valves are fully closed. Maintenance on equipment in operation should be limited to lubrication, packing adjustments, minor repair, or as allowed by the manufacturer's instructions.

3.8.3 Electrical. Lock out and tag main switches of electrical equipment before beginning work. Do not remove any tag without first checking with the person who initiated the tag. Any unusual motor temperature, noise, vibration, and so on should be reported and logged.

3.8.4 Underground Procedures. Where excavation of underground facilities is undertaken, OSHA-approved trenching procedures must be followed in the United States. Sloping or shoring requirements vary with local geologic and soil conditions. Requirements for sloping, benching, and shoring are found in OSHA 29CFR1926.652, appendices A, B, C, and D (OSHA 2002a). When excavating, the operators must be alert to both overhead power lines and underground utilities that pose a hazard to excavation equipment or personnel. Buried gas and electric lines pose the greatest hazard to operators, but accidents involving communication lines, water lines, and other utilities can be very costly to those causing the damage. Telephone numbers for all underground utility locators should be included in the operation and maintenance plan.

3.8.5 Other Safety Considerations. Workers entering structures must realize that surface drainage systems are likely to be part of the structure and subject to high flow conditions. Workers entering during these times should be protected by safety lines, flotation devices, and lighting. System features and signage designed to provide public safety should be kept in good condition. For further information, refer to OSHA safety requirements for workers in confined spaces (OSHA 2002b).

CHAPTER 4
WATER QUALITY

4.1 GENERAL

Periodic observation of flow is necessary to monitor possible water quality degradation. The potential for pollutants to be present constantly changes. Improvements and/or developments within the drainage basin can generate substances having a pollution potential that could be conveyed to the subsurface drainage system.

An increase in flow rate is often accompanied by higher velocity, which is more effective in transporting pollutants. This, in fact, could be the only cause for concern of such pollutants as suspended solids and/or turbidity. Quantity and quality should be monitored simultaneously.

USEPA and other local regulations affect the allowable water quality of subsurface drainage. The water quality requirements will affect the design of the system.

4.2 ENVIRONMENTAL INDICATORS

A review of the area should be performed to determine any changes since the construction of the subsurface system. These changes will then have to be evaluated as to possible effects on the subsurface flow. Water sampling of aquifers and watershed sources representing existing and potential sources of subsurface water supply may be required. Certain parameters and their background levels can be expected to occur naturally in the water due to the existing environment. By visual inspection or through personal observation, a determination can be made for the necessity and extent of a field sampling program. If test results show unusual concentrations or unexpected constituents in the water, further investigations could be necessary. A treatment program may need to be implemented, or modifications may need to be proposed that would mitigate or eliminate adverse effects caused by the problem constituents.

4.3 WATER QUALITY STANDARDS

Limitations may be imposed on the discharge based on the receiving waters. Water quality standards limit the concentration of various parameters to be discharged. Standards are generally established by federal, state, or local governments and are subject to continual revisions. These standards can be used as a basis for evaluating water quality. The suggested list of parameters for analysis include, as a minimum, temperature, color, odor, pH, dissolved oxygen, total suspended solids, and turbidity. Considerations include inorganic chemicals, heavy metals, corrosiveness, organic chemicals (pesticides/herbicides), and microbiological and radioactive materials.

Sampling points and frequency can best be determined in the field. Sampling points should be located at strategic points throughout the system.

An important reference on water quality standards is ASCE 33-09, *Comprehensive Transboundary Water Quality Management Agreement.*

CHAPTER 5
PERIODIC INSPECTION

5.1 GENERAL

Most systems experience long-term flow capacity reductions. Maximum subsurface drainage system performance requires maintenance programs that keep systems clean, soil tight, structurally intact, and free of debris.

Routine frequent inspections of all subsurface drainage systems are necessary to ensure the continuing level of water management intended by the initial planning and design. A uniform inspection schedule applicable to all sites cannot be established due to regional variations in climate or environment and local conditions affecting land usage. Experiences with similar systems may indicate specific local problem areas that should become a formal part of an inspection program.

Record keeping is important. Complete records of previous inspections provide a gauge for comparison to determine the rate and severity of deterioration.

5.2 UNDERGROUND SAFETY

All persons inspecting conduits below the ground surface or in any other confined space in the system should take the utmost caution when entering these areas. Deadly gases, such as methane, can be produced by decaying material within the storm drain or these gases can seep in from adjacent sewers or gas lines. All subsurface drains should be considered suspect even though they have numerous air inlets to the surface. In the United States, all persons entering underground conduits or other confined spaces are required to conform to OSHA regulations as found in 29CFR1910.146 (OSHA 2002b).

5.3 INSPECTION

Inspections should determine the current operating and structural status of the subsurface drainage system and provide thorough examination of symptoms of developing problems that can alter future performance. Typical elements comprising a thorough inspection are as follows:

- Check for accumulations of debris, rodents, or other flow impediments at inlets and outlets. Flap gates and other water control devices should operate freely.
- Inspect the system interior, if possible, for tree or other vegetation roots, mineral deposits, trash or silt accumulations, and other foreign objects obstructing flow paths. Often, televising the system is required to provide this service.
- Evaluate ground surfaces for evidence of subsurface drainage system leakage. Excessive groundwater from inoperable subsurface drainage systems may cause prematurely distressed pavements, loss of healthy vegetation, and topsoil saturations.
- Examine inlet and outlet areas for evidence of soil erosion, which generally leads to scour, undermining, and caving of adjacent soils supporting the subsurface drainage system. Soil erosion quickly leads to reduced structural and hydraulic performance.
- Inspect all visible structures such as catch basins, headwalls, culverts, and outlet pipes for signs of wear or breakage.
- Check upstream for evidence of backups or prolonged surface water presence that indicates reduced inflow. Check downstream for evidence of foreign materials that indicate reduced filtration of soils, diminished screening of foreign particles, or structural degradation of the drainage system itself. Visual inspection methods are frequently enhanced by electronic and optical aids such as television cameras, fiberoptic scopes, and laser-beam equipment. These inspection tools reveal cracks, displacements and misalignments, and other interior problems with minimal disturbance to the subsurface drainage system.

CHAPTER 6
MAINTENANCE

6.1 GENERAL

Maintenance of subsurface drainage facilities is often neglected. Most of the facilities are out of sight, so there is little public pressure to carry on maintenance programs. Failures tend to develop gradually, as opposed to being catastrophic, so that a failure may not be recognized for several years after it has occurred. However, costs of failure in terms of property damage and reconstruction can be enormous compared to the cost of routine preventive maintenance. Therefore, it is usually in the best interest of the owner to develop and implement an aggressive preventive maintenance program. Structures, such as manholes or outlet pipes, should be checked on a regular schedule for signs of structural distress and loss of hydraulic function. Appropriate repairs should be made on a timely basis.

6.2 CLEANING

Methods and schedules for cleaning depend on the type of subsurface drain being cleaned. The function of the drain, materials from which it is made, susceptibility to clogging, and type of clogging (roots, trash, chemical, or biological) all need to be considered. High-pressure hydraulic drain cleaners are satisfactory for most applications where access to the drain lines is provided. Initially, annual cleaning should be sufficient for roots and chemical or biological accumulations. As experience with the drain system is gained, the cleaning schedule should be adjusted accordingly.

Some installations, such as sport field drain grids or highway shoulder drains, may not provide access for mechanical cleaning.

Chemical treatment may be required in place of, or in conjunction with, mechanical cleaning. Chemical cleaning must be done in an environmentally responsible manner. For instance, iron ochre deposits, which can clog the openings in the pipe or restrict the carrying capacity of the pipe, sometimes require strong acid solutions for removal. These solutions must be contained until they are neutralized to the approximate pH of the receiving body of water.

6.3 MECHANICAL AND ELECTRICAL

Pumps and motors usually have manufacturer-recommended maintenance guides. When they do not, the design engineer should provide one, particularly if there are any unusual conditions. Appurtenances, such as covers, valves, or flap gates, may also require occasional attention. Flap gates that are removed for maintenance of the drain should always be immediately replaced. If frequent access is required, a different design of flap gate may be required to avoid losing the original one.

6.4 REPAIR

Repairs to subsurface drains should be made using original materials to the extent possible. Because subsurface drains are designed to last up to 100 years, original materials may not be available. When substitutions are made, care should be taken to ensure that the capacity of the drain is not diminished.

The same safety requirements should be in effect for repairs as for original construction. Maintenance crews must avoid the temptation to take shortcuts because of the short duration of their activities. Cave-ins and toxic fumes are even more prevalent in repair work than during original construction. Repairs to electrical equipment should be done under standard electrical safety rules.

6.5 REHABILITATION

Rehabilitation of a subsurface drain system may involve complete cleaning of all components; realigning portions of the system; replacing worn or outdated electrical components; replacing gates, valves, and other appurtenances; repairing or replacing major structures, such as manholes and outlets; replacing entire sections of drains; or providing additions to the original system. In general, rehabilitation should follow the same overall pattern as original construction, including engineering analyses, cost comparison, up-to-date materials selection, construction methods, and safety procedures.

3

CHAPTER 7
REFERENCES

Standards, including those applicable to these guidelines, are given in 1.1 of ANSI/ASCE/EWRI 12. Other references, including those applicable to these guidelines, are given in Chapter 10 of ANSI/ASCE/EWRI 12.

INDEX

Page numbers followed by *e*, *f*, and *t* indicate equations, figures, and tables, respectively.